Technology, Innovation and Industrial Economics: Institutionalist Perspectives

Essays in Honor of
William E. Cole

Technology, Innovation and Industrial Economics:

Institutionalist Perspectives

Essays in Honor of William E. Cole

Edited by

DILMUS D. JAMES
University of Texas at El Paso

and

JOHN W. MOGAB
Southwest Texas State University

SPRINGER SCIENCE+BUSINESS MEDIA, LLC

Library of Congress Cataloging-in-Publication Data

A C.I.P. Catalogue record for this book is available from
the Library of Congress.

ISBN 978-1-4613-7608-8 ISBN 978-1-4615-5697-8 (eBook)
DOI 10.1007/978-1-4615-5697-8

Printed on acid-free paper

This volume is dedicated to William E. Cole, a mentor, colleague and friend whose work inspired many of the ideas and insights contained in this collection and essays.

CONTENTS

CONTRIBUTORS

JOHN GROENEWEGEN is associate professor of economics at the Erasmus University Rotterdam, the Netherlands. He is fellow at the Tinbergen Institute and the Erasmus Institute of Philosophy and Economics. He is researcher at the Centre of French Studies of the University of Utrecht. He is past president of AFEE and currently general secretary of the European Association for Evolutionary Political Economy.

GEOFFREY M. HODGSON is University Lecturer in Economics at the Judge Institute of Management Studies, University of Cambridge. He has held visiting academic and research positions in Austria, Germany, Japan, Sweden and the United States. His publications include *Economics and Evolution* (1993), *Economics and Institutions* (1988) and *The Democratic Economy* (1984) and over 100 articles in academic books and journals.

DILMUS D. JAMES is Professor of Economics and Hanley Chair of Business Administration at the University of Texas at El Paso. After completing a BA and MA (1958) at the University of Texas at Austin, he earned a Ph.D. at Michigan State University (1970). He is past president of the North American Economics and Finance Association and of the Association of Evolutionary Economics. His research and writing have centered on technology capacity-building and science and technology policy in developing countries.

HANS E. JENSEN, Professor Emeritus of Economics at the University of Tennessee, Knoxville, received the Ph.D. from the University of Texas at Austin in 1961. His teaching and research interests lie in the history of economic though. In research, his emphasis is on inquiries into the origin and nature of economic paradigms and on investigations of the role of institutions in the work of the inventors of new theories. He has published articles and essays on the works of Clarence E. Ayres and Thorstein Veblen, John Maynard Keynes, Alfred Marshall, Karl Marx, John Stuart Mill, Joseph A. Schumpeter, and Adam Smith.

JOHN W. MOGAB, Associate Professor of Economics at Southwest Texas State University, received his Ph.D. from the University of Tennessee, Knoxville in 1981. His teaching and research interests are in comparative economics systems. He co-authored *The Economics of Total Quality Management* (Blackwell Publishers, 1995) and has published articles on Institutionalist thought and the continuous improvement firm.

STELLA S. SCHRAMM has worked both as a researcher and practitioner of economic development for 19 years. Her work has focused especially on technology development and its role in international competitive strategy. She received her Ph.D. in Economics from the University of Tennessee and served as doctoral fellow at the UN Institute for New Technologies, the Netherlands; research professor at the Center for the Research and Teaching of Economics (CIDE) in Mexico City and Director of the Economic Development Council/New Orleans & the River Region. She recently authored a report on the technology strategies available to industry clusters on behalf of the United Nations Industrial Development Organization in Vienna. Dr. Schramm is currently working for the Oak Ridge Chamber of Commerce to develop the economic development strategy for a national laboratory community.

STEVEN R. WILSON is an industrial economist in the Quality Standardization and Metrology branch of the United Nations Industrial Development Organization in Vienna. His current work focuses on the implications of standards and technical regulation for firms in developing countries. In addition, he is conducting research on the trade implications of standards and technical regulation for firms in developing countries. His most recent book *Beyond Quality: An Agenda for Improving Manufacturing in Developing Countries* (Aldershot: Edward Elgar, 1995) explores the application of continuous improvement technologies in developing countries.

PREFACE

Two motivations led to the publication of this book. The most obvious is our desire to recognize the valuable research contributions that William E. Cole has made to institutionalist economics, as well as his long career as a teacher at both the undergraduate and graduate levels. The second motive grew out of the character of Cole's institutionalist thinking and writing. Institutionalists, along with many other heterodox economists, have frequently been criticized for concentrating on a spirited critique of traditional, orthodox economics without proffering any concrete policy alternatives. The editors are compelled to admit that at times, this observation has been accurate. The editors are convinced, however, that the corpus of William Cole's work has centered on the positive insights that institutionalism can convey. The emphasis has been on how institutional economics can inform policy and the essays in this volume, to a very large extent, reflect this bent.

The original idea for the book came from Hans E. Jensen, one of the contributors to the volume. It was a capital idea and the editors express gratitude for his initial impetus and for guidance that he supplied in the beginning and throughout the project.

We are grateful as well to all of the other contributors who, despite the press of professional and personal obligations, provided their manuscripts, and editorial revisions, in a timely fashion. Contributors include former students of William Cole, a colleague from the University of Tennessee-Knoxville, and colleagues from a broader range of academia. The editors sincerely thank them all.

The editors' initial contact with Kluwer Academic Publishers was Zachary Rolnick, economics editor and his early enthusiasm was welcome indeed. The project proceeded under the direction of Ms. Ranak K. Jasani, acquisitions editor for economics, who showed great patience and was helpful at all stages of the project. The same can be said for Ms. Yana Lambert, editorial assistant, who provided detailed editorial advice. The editors express profound gratitude to each of them. A special word of appreciation goes to Ms. Sigrid Westphal who performed her magic in producing the manuscript despite her duties as Instructor for the Department of Marketing and Management at the University of Texas at El Paso.

Dilmus D. James
University of Texas at El Paso

John W. Mogab
Southwest Texas State University

1

INTRODUCTION

Dilmus D. James and John W. Mogab
University of Texas at El Paso and Southwest Texas State University

The impetus for writing this volume is to honor William E. Cole, who has been a mentor, colleague, and friend, on the occasion of his retirement. Throughout his career, Cole has contributed in many ways to the institutionalist school of thought. As an author, teacher, critic and member of professional associations, Cole has left his imprint on institutional economics. Each paper in this collection reflects an intellectual contribution he has made to institutionalism during his distinguished career. Taken together, the volume represents a contribution to the instititutionalist literature in the areas of technology, innovation, and industrial economics.

Although many readers are, undoubtedly, familiar with institutional or evolutionary economics, for the less initiated, we offer a brief exposition of some of the most salient tenets.[1] We will use "institutionalist" to connote Old Institutionalist Economics,[2] to distinguish it from a distant cousin, New Institutionalist Economics inspired by Ronald Coase and championed by Oliver Williamson. Often the terms "institutionalist" and "evolutionary" are employed interchangeably to describe the same group of economists, but we will, rather arbitrarily, stick with the former designation throughout our abbreviated discussion.

INSTITUTIONALISM: A COMPARISON WITH NEOCLASSICAL ECONOMICS

The institutionalist approach to economics can, perhaps, best be delineated by contrasting it with conventional neoclassical economic thought. Neoclassical economics remains heavily influenced by the physics of Isaac Newton while institutional economics owes more to Charles Darwin's evolutionary biology. Orthodox economics views both micro- and macroeconomic phenomena as inherently stable with strong equilibrating mechanisms operating through market forces. Institutionalists, as will become more readily appreciated below, see an evolving world where disequilibrium and continuous flux is the norm. A good deal of the explanation for the disparity lies in the differing assumptions regarding human behavior. In a neoclassical world rationality prevails in which individuals maximize satisfaction and institutions approach efficiency through a Darwinian/Spencerian weeding out of the less efficient institutional arrangements. The institutionalist world features much more complicated behavioral influences. Cultural values; mores; folk beliefs; and institutionally determined attitudes, tastes, and preferences, that are usually held as given in, or endogenous to, neoclassical economics, are enthusiastically embraced by institutionalists and are regarded as crucial to understanding how individuals and institutions behave. In addition, information asymmetries, bounded rationality and psychological dissonance are taken more seriously by institutionalists who, consequently, have far less confidence that unfettered markets will result in efficient institutional configurations. While recognizing that prices of goods, services and factors of production are important manifestations of information in a market economy, institutionalists insist that there is a far richer reservoir of important information available in social institutions and the corpus of technological knowledge that can guide the allocation of resources along more beneficial trajectories.

Further differences emerge, partly based on those already mentioned. Neoclassical economists are convinced that they have the answer (as do Marxists, albeit their answer is different) -- eliminate impediments to the operation of price-dictated free market allocation of productive and distributive activities and the economy can move closer to the best of all worlds, Pareto optimum. Furthermore, the beneficent dividends from the price mechanism is seen as universally applicable. The neoclassical economic approach is being urged for developing countries in Africa, Asia and Latin America as well as Eastern European countries in transition. Presumably Trobriand Islanders, Zuni Indian or Bantu tribes could all share in a market driven bonanza by letting the market work. Once again, most institutionalists do not deny that an efficient market system, itself an institutional arrangement, can be a useful and potent mechanism for directing economic activity, but they argue that progress in socioeconomic affairs is a far messier and uncertain undertaking. Institutional specificity or particularism must be taken into consideration and the process of problem solving must proceed through a healthy degree of trial-and-error and experimentation. Institutionalists, by and large, believe that the ability to make accurate forecasts through traditional analytical techniques is greatly overrated and, rather than moving teleologically toward some optimum, solutions to existing problems (not to mention random institutional and technological change) thrust up additional problems. For institutionalists, the exercise involves attempting to achieve some net gain in human welfare through measures that ordinarily cannot escape some readjustment as events unfold.

Institutionalists, who pride themselves on tackling practical, real-world problems are highly critical of the proclivity of neoclassical economists to create elaborate, abstract models that are long on elegance and internal consistency, but all too frequently short on relevance. Institutionalists routinely "trespass" into other disciplines by borrowing liberally from anthropology, history, political science, and sociology when they are believed to be useful in enriching economic analysis and, therefore, are critical of the narrow focus of neoclassical methodology. Institutionalists also reject the ahistorical nature of much neoclassical analysis--for them, history counts and small differences in initial conditions in technological capacity, institutional setting and physical environment for individuals, enterprises, industries, industrial clusters, or national economies affect the character of evolutionary change. Lock-in or path-dependent change, cumulative causation and irreversibilities affect the pace and direction of economic change. The neoclassical practice of stipulating conditions that will move an economic entity from one equilibrium to another in a frictionless manner and, with a reversal of conditions, move back to the original position is, according to institutionalists, an exercise more firmly grounded on fantasy that realism. Partly because the world is complex and inherently unstable, manifesting changes along unpredictable trajectories, institutions and technologies must be scrutinized and judged more thoroughly and deliberately than the market mechanism is capable of accomplishing. Institutional economics is avowedly valuational as opposed to the ostensible value-free analytical mode with which neoclassical economists feel most comfortable.

Illustratively, two further differences are instructive, namely the treatment of power and technology. Neoclassical economists' preoccupation with power revolves primarily around interventions (the state and labor organizations are high on the list) that interfere with price-guided allocation of resources based on the unimpeded operation of markets. Without such interventions, they believe a market-driven, price-directed economy can diffuse power sufficiently to prevent privileged groups from exercising unwarranted influence or control over others. For institutionalists, power asymmetries are extremely prevalent. Indeed, the existence of power asymmetry is virtually a definitional characteristic of an institution. Property rights, the family, the military, religion, the nation state, the corporation, the news media, and so on, are receptacles of institutional power that inevitably exercise some sway over our behavior. While they are concerned with monopoly power, institutionalists view power asymmetries as ingrained in many more aspects of human relations. One primary policy objective is to distribute and steer the exercise of power away from invidious, status-based actions toward more productive pursuits.

Until very recently, mainstream economics treated technological change as either being generated exogenously or as subject to cost/benefit analysis (similar to ordinary investment decisions). The starker textbook renditions assume all firms are capable of adopting new technologies instantaneously and simultaneously throughout an industry. Institutionalists, on the other hand, are intrigued by the root causes of scientific and technological change. Especially in the Thorstein Veblen-Clarence Ayres axis of institutionalism, tool-using activities (including both physical and conceptual tools) interact with ceremonial aspects of human behavior in ways that vitally affect the pace and direction of technological change. Ordinarily the ceremonial or institutional manifestation of human behavior tends to be change resistant, i.e., supportive of the status quo, whereas technological change will virtually always result in altered institutional patterns. Furthermore, the generation of new technologies is endogonized into institutionalists' analysis and the acquisition and absorption of technology are time and resource consuming activities.

This summary of orthodox economics focuses exclusively on the received core and thus is something of a caricature of neoclassicism. Recently there has been a healthy amount of diversity and probing around the edges by economists trained in traditional economics. The possibility of multiple equilibria, the new (or endogenous) growth theory, the admission that inefficient institutions can evolve and be perpetuated, and the recognition of the importance of information asymmetries are but a few examples that have resulted from the work of such mavericks. Nonetheless, the chasm between neoclassical and institutional economics remains sizable.

WILLIAM E. COLE'S PROFESSIONAL CAREER

William E. Cole's professional career has been primarily devoted to academic endeavors, with periodic sojourns into the realm of international agencies. He prepared for his professional career by attending The University of Texas at Austin where, in 1952, he earned a BA (with honors) in Government. After a brief stint in the US Army, he joined Procter and Gamble as an internal auditor, was promoted to credit manager, and later became a Supervisor of Office Services. He returned to The University of Texas at Austin to pursue the Ph.D. degree in Economics, which was awarded in 1965. His Ph.D. studies were concentrated in the fields of Economic Development and Latin American Studies. While working toward the Ph.D. degree, he also studied at the Instituto Tecnologico y Estudios Superiores de Monterrey in 1963, working with a team to construct a computerized decision making model of the steel industry in Mexico. The following year he returned to Mexico as a Fulbright-Hays Fellow to continue his research on the Mexican steel industry. His studies of the Mexican steel industry formed the basis of his doctoral dissertation, which was published by the University of Texas Press in 1967. While at The University of Texas, his teaching career was started with his assignment as a Teaching Assistant from 1964-65.

While finishing his doctoral dissertation, Cole joined the faculty of St. Mary's University in San Antonio as an Assistant Professor, teaching a graduate seminar in Latin American development. Upon completion of the Ph.D. he took a tenure-track position with the economics faculty at The University of Tennessee, Knoxville. While at Tennessee he continuously refined his considerable teaching skills at both the undergraduate (Principles of Economics, Economic Development, Latin American Economies, and International Economics) and graduate (Economic Development, and Latin American Economies) levels, for which he was recognized with Outstanding Teaching Awards in 1974 and 1991. At Tennessee he also continued to conduct research in the fields of Latin American studies and economic development, returning to Mexico often during the 1960s on various research projects. His teaching and research efforts were rewarded in 1969 when he was promoted to the rank of Associate Professor of Economics.

During 1970-72 he took a leave of absence from The University of Tennessee to join the United Nations Industrial Development Organization as an Industrial Development Officer. In this capacity he directed preparatory work on, and headed, a

World Bank Industrial Development Mission to Kenya and Uganda; oversaw an Industrial Team of the Preinvestment Studies Mission of the World Bank which conducted economic surveys of the countries in the East African Community; conducted research in preparation for, and led, an Industry Team for the International Labor Office's World Employment Mission to Kenya; and represented the United Nations Industrial Development Organization at the International Labor Office's Conference in Kenya.

In 1972, he returned to The University of Tennessee to resume his teaching duties and his research on Mexican economic development. His teaching and research accomplishments once again were rewarded when, in 1974, he was promoted to the rank of Professor of Economics. He continued in this position until 1983 when he took up the reins of administration, accepting the position of Head of the Department of Economics. He served as Department Head until 1986, combining the administrative duties with his teaching and research responsibilities.

By the late 1980s, his research interests were expanding to include other Latin American countries and regions of the world, as well as new fields of study. In 1987, he represented The University of Tennessee in an employee orientation program of the Saturn Corporation. This experience, in conjunction with his newly developing research cooperation with the Management Development Center (MDC) of The University of Tennessee, led him to shift his research agenda toward the topics of the continuous improvement firm and total quality. During 1987-89 he served as the Principal Investigator on a joint grant with the MDC to develop a business and economic research center at the Federal University of Ceara, Brazil that was funded by the Tinker Foundation. In 1988-89, he also was the Principal Administrator on a contract with the Brazilian government to develop and deliver a productivity and quality-training course for Brazilian government, business and university personnel. In 1990, he continued his work in the field of total quality as the Principal Administrator on a contract with UNIDO to deliver a seminar on Total Quality Management in Sao Paulo, Brazil. During that same year, he assisted the State Planning Commission of the People's Republic of China under the auspices of the United Nations Development Program, and, at a symposium sponsored by the Foreign Ministry of Japan, delivered the keynote speech on "Development Problems and Foreign Assistance Prospects for the 1990s."

In 1996, Cole retired from The University of Tennessee and his many years of outstanding service to the university were recognized by awarding him the rank of Professor Emeritus of Economics. He did not, however, give up his lifelong work. He founded Tenn-Tex Associates and, as Senior Partner, continues to conduct research and work as a consultant on the topics of continuous improvement and total quality in developing countries.

WILLIAM E. COLE: A SYNOPSIS OF RESEARCH

The research and writings of William E. Cole have been far ranging. He began with a book on the Mexican steel industry (1967) and since has investigated (1) dualism in labor markets and the informal sector in developing countries (Cole and Sanders 1976; Cole and Fayissa 1991), (2) migration in the Third World (Cole and Sanders 1983a, 1983b, 1985; Cole 1996), (3) Mexican agriculture (Cole and Sanders 1972, 1974; Cole 1990), (4) technological and institutional influences on decision making by Mexican authorities involved in large-scale river valley development (Cole et al. 1984; Cole and Mogab, 1987) and most recently (5) the economic implications of different managerial styles in the manufacturing sector, especially as to how they impact on innovation and competitiveness (Stephens and Cole 1988; Cole 1991; Bos and Cole 1994; Cole and Mogab 1995).

While William Cole is an institutionalist, he is anything but doctrinaire. He and John Mogab (1987) raised serious questions about the Veblen-Ayres dichotomy between technology and ceremonialism with their comparison of the accomplishments of two dependencies of Mexico's Ministry of Hydraulic Resources, the Papaloapan Commission and the Tepalcatepec Commission. Despite the fact that the Papaloapan Commission had a far more technically competent staff and that the Tepalcatepec Commission was more highly politicized, the latter was able to get more done. There is a strong suggestion that institutionalists may be too Euclidean in distinguishing technological and ceremonial aspects of human behavior. In another example, Cole (1990), after a review of the history of Mexican agriculture and the institutional setting in which the Mexican ejido sector operated, stated that "It may be a propitious time to conclude that the intertemporal transfer of the ejido as an institutional form has not succeeded (123)." He goes on to recommend that the ejido, long the darling of heterodox economists, be gradually transformed to privately-owned smallholder units.

Cole's work on internal migration in developing countries, most of which was done in collaboration with Richard D. Sanders and featured the migration experience of Mexico, offers several significant contributions. Some telling criticisms were leveled at the Todaro neoclassical model of migration in which rural-urban wage differentials and the probability of obtaining a job in an urban setting are the fundamental "demand-pull" factors determining the volume of emigration from the countryside to cities. Cole and Sanders contend that this model, the dominant paradigm in the field, drastically discounts "push factors," namely population growth in a situation where land is limited. The usual assumption was that, while low income was a problem in the rural sector, securing a workplace was not. Cole and Sanders demonstrate that under certain institutional environments, individuals, especially women, may literally have no place to go within the rural sector.

Furthermore, while the Todaro scenario has arriving migrants queuing up for jobs in the modern sector, the Cole and Sanders rendition has the vast majority being rather quickly and facilely absorbed into the urban informal sector where there are few barriers to entry, low skill requirements and minimal capital costs per job. This means that rapid rural-urban migration is not the unmitigated disaster than is usually supposed since earnings in the informal-urban occupations, meager though it may be, beats having no earning opportunities at all in the rural sector. Policy suggestion are proffered for attracting rural migrants to secondary cities to alleviate the pressures felt by megametropolises and for upgrading economic conditions in the informal-urban activities. The manner in which institutional specificity can be of overwhelming importance is illustrated by Cole's analysis of the differences in institutions related to the inheritance of land in Mexico and India (Cole 1996). He concludes that the Mexican practice of leaving land to the eldest son, contrasted with the India tradition of dividing of land equally among children means that land reform in Mexico is likely to be less effective in providing sufficient levels of productive employment in agriculture.

In a more recent publication, Cole and Mogab (1995) contrast the mass producing firm (MPF) and continuous improvement manufacturing firm (CIF). The mass producer, in accordance with traditional economic theory, has the objective of maximizing profit for the benefit of top management and shareholders. Conditions that are conducive to the success of mass production techniques are long production runs and accurate prediction of product demand. Because of the overriding goal of maximizing profits, and very often, because of the financial environment in which short-run profits are stressed, labor is treated as a commodity easily hired or laid off. Furthermore, unit costs dominate subcontracting arrangements, managerial positions are frequently filled by professionals from outside the enterprise, and employees tend to be specialized in task-specific assignments. On the other hand, the continuous improvement firm (CIF) strives to increase customer value, treats labor as a valuable asset, cultivates lasting symbiotic relations with supplier firms, more frequently enjoys the luxury of focusing on long-term development, and routinely rotates employees among different functional departments.

The disparity in how labor is viewed means that in case of bad times, the MPF has no hesitation in downsizing its operations and reducing its workforce. The continuous improvement producer, conversely, will go to great lengths to retain its worker corps intact. The MPF, thus, faces a greater incidence of labor mobility and this results in less incentive for firm-sponsored training; the opposite is true for the CIF.[3]

Information tends to be made available to all in the CIF, whereas in the MPF it moves in compartmentalized vertical "silos" circumscribed by departmental designations (research and development, production, personnel, marketing, and so on). For the CIF, worker suggestions for improvements originate, for the most part, on the shop floor, and constitute the paramount source for quality-upgrading and cost-reducing innovations, a situation which differs markedly from the MPF's heavier reliance on in-house or external R&D leading to larger, more discreet and more expensive alterations in production technique. As a result, technical improvements, including improvement in product quality, by the MPF are evaluated by a standard textbook comparison of marginal cost and marginal revenue, whereas the CIF reaps the benefits of innovations, including quality control, that flow routinely from the enterprise's productive undertakings. In addition, the distinction between short run and long run is maintained for the MPF, but are blurred considerably for the CIF where capacity-stretching innovations mean that output can be expanded even though the basic plant hardware remains unchanged.

The CIF is extremely flexible and adept in responding to changes in demand, a facility that is uncharacteristic of the MPF. The CIF model, especially in the motor vehicle and electronics industries, has become dominant in international competition. The CIF managerial style evolved over an extended period of time and there are significant difficulties facing MPFs that attempt to compete with their largely outmoded approach. But making the transition, whether partial or complete, to a CIF will create intra-firm resistance and incur substantial costs.

Cole and Mogab (1997) question the neoclassical view on the role of specialization in productivity growth. Since Adam Smith's *Wealth of Nations* it has been conventional in economics to assert that specialization (or division of labor) is the ultimate source of the increased productivity associated with economies of scale. Cole and Mogab argue that the role of specialization in the firm and the tradition in economics of associating it with division of labor should be re-evaluated because of the recent evolution of the CIF. They compare the Smithian view of endogenous technological change with the institutionalist approach (as expressed in works by Ayres, DeGregori, Ogburn and Gilfillen). In expanding upon the institutionalist approach they formulate a tentative theory of endogenous technological change and conclude that the CIF combines division of labor with a generalization of the work force to achieve endogenous technological change and, thereby, increases productivity.

A Preview of the Essays

The following three chapters by John Mogab, Stella Schramm and Steve Wilson are extensions of the Cole inspired work on the CIF. In chapter two Mogab makes the case that Japan's regulations that inhibit entry into many of her markets may not be predicated on old-style protection from foreign competition as is widely believed by western nations. After reviewing the theory of contestable markets and the characteristics of the CIF, Mogab concludes that Japan may be acting to preserve an economic environment necessary for CIFs to expand. Indeed, the regulations are probably intended to check competition from potential domestic entrants into sheltered markets rather that foreign rivals, since the latter, at least in the short run, would find it difficult to match the CIF's efficiency and innovativeness, as well as the consumer value their products offer the public. An infusion of competition from either new entrants in Japan or from abroad would increase static efficiency, but would undermine the prospects for steady growth that a CIF demands in order to offer secure employment, seniority wages and good prospects for promotion which are the necessary incentives for worker-derived innovations and more cooperative management-labor relations. Apparently, Japan is willing to sacrifice short-term benefits from competition in return for the dynamic gains inherent in the CIF's management style and mode of operation. The author, thus, disagrees with western observers who believe that Japan is moving toward a free market system that will include opening up to foreign penetration of her markets. He concludes that this view ignores the institutional environment within which Japan's regulations are being implemented.

In the following chapter, Schramm looks inside the CIF, asking the question: What institutional features allow the CIF to continuously innovate? As a prelude to answering the question, Schramm reviews the treatment of technology in economic analysis. She argues that the conventional cost/benefit analysis used to evaluate exogenously produced technological change (i.e., investment) was workable when the rate of technological change was relatively slow and the organization of technological development was relatively universal. With the recent evolution of the CIF and its concurrent approach to continuous technological development, Schramm argues that the conventional approach is becoming less useful. An alternative approach is sketched beginning with an analysis of the components necessary for an organization to achieve continuous improvement. The author suggests that a firm must be able to implement simultaneously adequate controls to ensure the performance standards are met (or exceeded) and provide adequate support for continuous change. To do so, it is argued, controls need to be diffused among the labor force to ensure competence and authority to institute change within those most actively involved in the productive activities. The author concludes that the CIF is able to innovate continuously because it is characterized by diffuse management and decision making, superb coordination of activities, articulated methods of innovation, and high levels of discipline in worker behavior.

Chapter three by Steve Wilson applies the lessons derived from analyzing the CIF to industrial districts. An industrial district is a geographic agglomeration in which an industry, usually dominated by small-and medium-sized firms, manages to achieve a peculiar balance between fierce competition in some activities and cooperation in other realms. Much work on small-scale industry has been done, but Wilson contends that there has been a dearth of analytical investigations. Much of the literature on small-scale production in general, and industrial districts specifically, is reviewed. In addition, he lays out what he considers to be the fundamental characteristics of industrial districts and the conditions leading to "collective efficiency," a term that is more comprehensive than external economies as they are usually conceived. Wilson points out that studies of industrial districts have failed to uncover whether improvements continuously and routinely arise from production or whether they are due to discrete improvements that originate in isolation and are then tried on the shop floor. This, Wilson believes, is crucial to determining the degree of "embeddedness" of continuous improvement in the industrial district model which, in turn, will determine the sustainability of industrial districts as competitive entities in the global economy.

The Japanese firm is also the focus of chapter four wherein John Groenewegen analyses the Japanese enterprise "as a dynamic institution in which routines, competence and strategy are central." Here old institutional economics meets new institutional economics as Groenewegen, after reviewing the traditional or neoclassical theory of the firm, looks at transaction cost economics in the same context. Although he finds the transactions cost approach still largely rooted in neoclassical economics, the author uses insights from the new institutional economics that are useful for constructing a dynamic theory of the firm. The author proceeds to develop his dynamic theory that goes well beyond viewing an enterprise as a production function (a la neoclassical economics) or a bundle of contractual arrangements (a la transaction cost economics). A firm's strategy, routines, and tendency to be path dependent are all crucial characteristics, but they cannot be properly understood by observing the enterprise in isolation. The general institutional environment, the role of the state and subcontracting relations all impinge on firms operating in the Japanese economic setting. Particular emphasis is placed on the economic, organizational and financial interactions among industrial enterprises, horizontally structured groups of enterprises (kinyu keiretsu), the general trading company, and the group (or main) bank. Although the dynamic theory of the firm is the cornerstone of the chapter, the author concludes that the neoclassical, transaction cost and dynamic approaches are complementary with their contributions hinging on the specific context in which the firm is to be analyzed.

Cole's interest in innovation within the firm is reflected in Jensen's essay on the role of the entrepreneur in Adam Smith's writings. In chapter five, Hans Jensen finds a far richer and active Smithian entrepreneur than is conventionally ascribed in the field of history of economic thought. In his examination of Smith's ideas, Jensen widens the notion of entrepreneurship to include some activities that are not ordinarily considered entrepreneurial (thus the "opportunistic interpretation" in the title). After reviewing briefly the goals that Smith sought to achieve by writing the *Wealth of Nations*, the author provides an exposition of Smith's ideas on invention and a detailed exposition of Smith's writing on innovation. Smith believed that there would be a decline in work-place inventions and a corresponding rise in inventions by more educated professionals. Jensen believes that Smith's thinking regarding the interaction of inventors and innovating entrepreneurs can be regarded as precursors to many of the ideas expressed in this volume. He concludes with several reasons for believing that contemporary students of economics and entrepreneurship can benefit from revisiting the *Wealth of Nations*.

Since William E. Cole had a career-long interest in Latin American economic affairs, it is fitting that one essay focuses on the region. Dilmus D. James, in chapter six, asserts that science and technology were never close to the center of policy in either the structuralist or neoliberal strategies for development. He believes that the objective of technology capacity-building is a crucial ingredient in any viable strategy for sustained socioeconomic development and that the neostructuralists, an influential group of economists in the region, are well positioned to give it the attention that it deserves. After briefly highlighting some inadequacies of science, technology and innovation in Latin America, James characterizes the neostructural school of thought as attempting to combine, in a pragmatic fashion, the best, most workable components of the older structuralist state-dominated, inward-directed path to development, and the more recent trials of market-driven, outward-oriented neoliberalist regimes.

Neostructuralists should, the author contends, feel reasonably comfortable with incorporating science, technology and innovation into their scheme of things for several reasons. First, despite inadequacies in scientific, technological and innovational activities in Latin America, there is a rich profusion of literature on the topic. The chapter provides a glimpse at the types of studies that are available. Many neostructuralists are familiar with this literature and, indeed, a few of them have contributed to it. Thus, a convenient entree into the economic consequences of science and technology policy in Latin America is at hand. Second, neostructuralists' preoccupation with macroeconomic policy should represent no problem because there are palpable interactions between the condition of a nation's economy and the ability to foster technology capacity-building. Third, neostructuralists have always regarded human resource development as fundamental for socioeconomic progress;

the interface between education and technology capacity-building offers a fertile area for neostructuralist exploration. Finally, the neostructuralists see a positive role for the state which complements the notion that technology capacity-building can be advanced through judiciously selected measures in which the state can be instrumental.

The volume concludes with Geoffrey Hodgson's interesting and provocative comparison of American and European institutionalism. Institutional economics is currently enjoying a revival and is gaining a distinctive identity in Europe. Hodgson claims that the two versions, while sharing a core set of concepts and approaches, may have begun to diverge. Although both schools have roots in the writing of Thorstein Veblen and John Commons, the additional influence of Karl Marx is much stronger in the European arena. Furthermore, while post-1945 American institutionalism was strongly influenced by Clarence Ayres, as yet he has had little influence in Europe. He analyzes the reasons why American-style institutionalism failed to influence European social scientists significantly during the twentieth century and goes on to identify specific differences between the two schools of thought. These differences are evident in work on the nature of institutions, and on valuation and welfare. By applying the insights of institutionalism to the evolution of institutionalism itself, Hodgson concludes that cumulative and path-dependent processes should lead to the expectation of further divergences between European and American institutionalism.

ENDNOTES

[1] Two excellent sources on institutional economics are the "Foundation Papers" in the September and December 1987 (Vol. 21) issues of *Journal of Economic Issues*, and Hodgson, Samuels and Tool (1994).

[2] The seat of Old Institutional Economics is the Association for Evolutionary Economics which sponsors the *Journal of Economic Issues*. More recently, the Association for Institutional Thought and the European Association for Evolutionary Political Economy have made important contributions to institutionalist thought.

[3] Since much technical learning is firm specific, the same contrast applies to worker incentives to invest in their own training.

REFERENCES

Bos, Antonio and William E. Cole. "Management Systems as Technology: Japanese, US and National Firms in the Brazilian Electronic Sector." *World Development* 22: 2(February 1994): 225-236.

Cole, William E. *Steel and Economic Growth in Mexico.* Austin: The University of Texas Press, 1967.

________. "Investment in Nutrition as a Factor in the Economic Growth of Developing Countries." *Land Economics* 47: 2(May 1971): 139-149.

__________. "Technology, Ceremonies, and Institutional Appropriateness: Historical Origins of Mexico's Agrarian Crisis." In James L. Dietz and Dilmus D. James, eds., *Progress Toward Development in Latin America: From Prebisch to Technological Autonomy.* Boulder, CO: Lynne Rienner, 1990, pp. 103-129.

________. "Competitive Economies and the Economics of Competition." In Michael J. Stahl and Gregory M. Bounds, eds., *Competing Globally Through Customer Value: The Management of Strategic Suprasystems.* New York: Quorum Books, 1991, pp. 14-31.

________. "Labor Migration and Urban Employment in Developing Countries: The Impact of Population Growth and Property Institutions." In John Adams and Anthony Scaperlanda, eds., *The Institutional Economics of the International Economy.* Boston: Kluwer Academic Publishers, 1996, pp. 161-177.

Cole, William E. and Bichaka Fayissa. "The Urban Subsistence Labor Force: Toward a Policy-Oriented and Empirically Accessible Taxonomy." *World Development* 19: 7(July 1991): 799-789.

Cole, William E. and John W. Mogab. "The Transfer of Soft Technologies to Less-Developed Countries: Some Implications for the Technology/Ceremony Dichotomy." *Journal of Economic Issues* 21: 1 (March 1987): 309-320.

________. *The Economics of Total Quality Management: Clashing Paradigms in the Global Market.* Cambridge, Massachusetts: Blackwell, 1995.

Cole, William E.; Steven M. Neuse and Richard D. Sanders. "TVA: An International Administrative Example." *Public Administration Quarterly* 8: 2(Summer 1984): 166-183.

Cole, William E. and Richard D. Sanders. "Income Distribution, Profits and Saving in the Recent Economic Experience of Mexico." *Inter-American Economic Affairs* 24: 2 (Autumn 1970): 49-63.

________. "A Modified Dualism Model for Latin American Economies." *Journal of Developing Areas* 6: (January 1972): 185-198.

________. "The Impact of Agriculture's Domestic Terms of Trade." *Growth and Change* 5: 3 (July 1974): 36-40.

________. "Population Growth and Employment: Mexico's Past and Kenya's Future." *African Studies Review* 19: 1 (April 1976): 151-163.

________. "Enfoque Econometrico del Modelo de Todaro sobre la Migracion Rural-Urbana: El Caso de Mexico." In Victor L. Urquidi and Saul Trejo Reyes, eds., *Recursos Humanos, Empleo y Desarrollo en la America Latina*. Mexico, D.F.: Fondo de Cultura Economica, 1983, pp. 173-192.

________. "Interstate Migration in Mexico: Variations on the Todaro Theme." *Journal of Development Economics* 12(1983): 341-354.

________. "Internal Migration and Urban Employment in the Third World." *American Economic Review* 75: 3 (June 1985): 481-494.

Hodgson, Geoffrey M.; Warren J. Samuels and Mark R. Tool (eds.) *The Elgar Companion to Institutional and Evolutionary Economics*, 2 Vols. Aldershot, England: Edward Elgar, 1994.

Mogab, John W. and William E. Cole. "Information and Specialization in Endogenous Technological Change." *Southwestern Journal of Economics* 2:1(April 1997): 79-98.

Stephens, Mark and William E. Cole. "The Brazilian Motor Vehicle Industry: A Holistic Approach to Project Evaluation." *Journal of Economic Issues* 22:1 (March 1988): 381-388.

2

INDUSTRIAL POLICY AND PROTECTIONISM IN POST-WWII JAPAN

John W. Mogab
Southwest Texas State University

In recent years Western government officials and business leaders have been not so quietly telling Japan to open its markets to foreign competition. None has complained more loudly than the U.S., as witnessed by the recent threat to impose a 100 percent tax on the importation of Japanese-made luxury cars and the Kodak versus Fuji film case filed by the U.S. against Japan in the World Trade Court. Westerners, indoctrinated in the principles of free market theory, generally believe that Japan's protectionist policy is nothing more than another misguided case of mercantilism. It is argued that protectionist policies not only harm foreign firms by preventing them from entering the market, but also cause Japanese consumers to pay higher prices for the goods. The overall result is an inefficient allocation of resources and a lower economic growth rate for Japan.

The Japanese government officials and business leaders, however, do not see it this way. The Japanese, having not been indoctrinated in Western economic theory, view regulation from a pragmatic perspective. Typically, they establish certain goals and then put in place laws and rules (regulations) that they believe will achieve those goals. Seen from this perspective, the goal during the immediate post-WWII period was to achieve industrial development. A variety of government regulations were imposed with one outcome being the evolution of a new type of industrial firm, the continuous improvement firm (CIF). The Japanese CIFs became formidable competitors in the global industrial markets and, by the early 1980s, they heralded the arrival of Japan as a leading economic power. The global presence of Japan's CIFs created new opportunities for the firms and altered the balance of power between the regulators and the firms. In light of the changing environment, the government regulators established a new set of goals and correspondingly changed its regulations. The new goals are to support Japan's international competitiveness and avoid a "hollowing out" of Japan's economy (Schaede 1994, 1-3). The corresponding changes in regulations have resulted in less use of across-the-board direct regulations, and more reliance on the use of administrative guidance and situational regulations. These changes in policy have given some Western observers hope that Japan is moving toward a free market economy and, thus, will open its markets to foreign competitors. "The Japanese economy, the argument goes, will soon be driven by free market forces only, and the economic system will become much like that of the United States" (Schaede 1994,1).[1]

It is argued in this paper that such a conclusion is unfounded, and is based on the application of Western economic theory without adequately taking into account the institutional context within which the policy is being implemented. Within the Japanese economy, CIFs are at the core of the manufacturing sector and public/private sectoral planning determines the form that competition takes. Given these institutional characteristics, the consequences of introducing free markets makes it highly unlikely that Japan will pursue such a course. The more likely course is that Japan's industrial markets will continue to be less directly regulated, but will not be characterized by free markets, as many Westerners apparently desire. They will, nonetheless, become more contestable by domestic CIFs. Contestable markets theory implies, however, that potential entrants must have access to the same technology as the incumbent firms. Because Japan's CIFs employ technologies that are internally developed and improved, potential entrants would need to be organized to do the same. Unfortunately most potential foreign competitors are not organized as CIFs and will, therefore, continue to be shut out of Japan's markets. A significant difference, however, will be that most potential foreign competitors will be shut out by their inability to duplicate the technologies of the incumbent CIFs rather than by direct government regulations. Furthermore, should market forces cause significant structural and institutional changes in Japan's CIFs that could threaten the strategic competitive advantage they hold in their domestic marketplace, it is likely that the policy will once again move toward greater regulation.

The analysis that follows is organized in four sections. The first section looks at the relationship between Japan's industrial policy during the post-WWII period and the evolution of the CIF. This is followed by a brief review of Western economic theory, including the theory of perfectly competitive and contestable markets theory. The third section is an analysis of the attributes that would likely be necessary to compete in the Japanese manufacturing sector. The paper concludes by laying out the future implications of Japan's post-1980s industrial policy.

JAPAN'S POST-WWII INDUSTRIAL POLICY AND THE EVOLUTION OF THE CIF

The foundation for Japan's post-WWII industrial policy was laid during the occupation period (1945-52) when public policy in Japan was directed by the Supreme Commander for the Allied Powers (SCAP). SCAP's goals for the occupation policy were, first and foremost, to destroy the military superstructure, and only secondarily to implement economic and social reforms. Of particular interest here are two economic reforms begun during the occupation period: 1) the attempt to reverse industrial concentration, and 2) labor reforms.

Industrial Concentration

A number of steps were taken to reverse the trend in industrial concentration in Japan. The most important was the dissolution of the zaibatsu, the family-controlled financial and industrial conglomerates that had dominated Japan's economy during the inter-war period. The Holding Company Liquidation Commission, set up in 1946, ordered 83 holding companies and 57 zaibatsu families to surrender their corporate shareholdings, dissolving the holding companies. The companies which had been under their control became independent firms. In addition, all cartel and control organizations that had been set up under the 1931 Major Industries Control Law were liquidated. These changes were combined with the implementation of the Antitrust Act of 1947, and the creation of the Fair Trade Commission to administer the antitrust laws. In 1948 the Board of Smaller Enterprises also was set up to promote the growth of small independent companies. "[T]he intention of SCAP was to decentralize the Japanese economy and reconstruct it according to the theoretical vision of perfect competition that guided American antitrust policy" (Best 1990, 174).

Despite the apparent intentions of SCAP, the policy of industrial decentralization did not last long in Japan. A revised antitrust law was implemented in 1949 that allowed companies to hold shares of stock in other companies. As a result, there evolved a new type of industrial network, called the keiretsu, with cross stockholdings among member firms. The keiretsu's typical organizational hierarchy consists of a "parent" company at the top, several primary subcontractors beneath it, followed by numerous lesser subcontractors and "commoner" companies (Sakai 1990,. 42). The keiretsu, in almost every case, centers on a main bank or a trading company as the parent company. The primary subcontractors are large manufacturing firms, e.g., Mitsubishi Electric, Sumitomo Heavy Industries, and Nihon Cement. The numerous lesser subcontractors and commoner companies are smaller firms that produce only a limited number of products or components (in some cases only one), with anywhere from a few workers to a few hundred employees. An important feature of the keiretsu is the maintenance of close long-term relationships among the member firms. The long-term, close nature of such relationships also carries over to the relationships between firm and customers. Kagano, et al., state, "Japanese firms emphasize 'relationship' itself and try to build and maintain good relationships with exchange partners even when the relationships create more constraints than benefits (Kagano et al. 1984, 35)."

Also during 1949, SCAP resurrected the prewar industrial policy agency, thereafter called the Ministry of International Trade and Industry (MITI). MITI has since been at the center of much of Japan's industrial policy. Schaede (1994) argues that post-WWII industrial policy in Japan can be divided into three phases.

"The first period, beginning in the mid-50s and ending with the first oil shock in the early 1970s, is characterized by extensive regulation, which aimed at economic recovery, development and growth. The second period, stretching over most of the 1970s until the second oil shock in 1979/80, is largely a period of flux. It is characterized by large-scale economic reorganization and industrial restructuring, towards increased exports and diminished dependence on other countries, in particular in terms of supply of raw materials. This period constitutes a mix of strict government guidance and regulation with increasing 'emancipation' of large Japanese firms that turned global, at a time of 'internationalization' in terms of flexible exchange rate and efforts towards enhanced world trade through GATT and other political forces. The third period, beginning in the mid-1980s and extending into the present, is also labeled in Japan the 'period of low, stable growth'. In this period, large firms turned increasingly 'international', and government intent with regulation shifted towards supporting their international competitiveness, while at the same time deregulating domestic markets in order to avoid a 'hollowing out' in the sense that Japanese firms transfer their main activities to other countries where regulation is less stringent." (Schaede 1994, 2-3) The first phase of industrial policy included the use of laws, rules and guidance to prescribe what kind of activities could be performed[2] and resulted in government dominance and control of all economic activities. Best's study of Japan's industrial policy found that "by the early 1950s MITI's formal industrial policy-making instruments included access to foreign exchange, foreign capital, imports, technology, and licenses, access to domestic credit and selective tax measures and authorization to create cartels." (Schaede 1994, 177)

The second phase was a transition period for Japan's industrial policy during which the power of large Japanese firms increased vis-à-vis government regulators. The shift in power ushered in the third period, beginning in the early to mid-1980s with a relaxation in domestic regulation and a shift toward indirect (rather than direct) regulation. Rather than prescribing a priori what could and could not be done, the government began to react to new business practices. Industrial policy concentrated on specific areas, taking into account industry life cycles. Japanese regulators increasingly relied "on situational regulation and administrative guidance and the carrot-and-stick mechanism as enforced by discretionary power of business licenses and permits. Also, the traditional mechanism of government-business relations, such as guidance through public corporations or promotion firms, or the amukadari, retired bureaucrats embedded in an 'Old Boy' network, are enforced to compensate for the decline in direct regulation" (Schaede 1994, 6).

MITI, however, has consistently used its industrial policy instruments throughout the entire post-WWII period within a joint public-private arena to articulate and coordinate long-term strategic planning at the sectoral level. Strategic sectoral planning first strives to articulate a long-term vision which suggests what the future sectoral contours will look like on a global scale. This is followed by an analysis of where Japanese firms fit within the global picture. Then MITI identifies policies to try to shape the strategies of individual firms to make the transition from the present organizational structure to those implied in the sectoral strategy. Best states, "The purpose of a sector strategy is to negotiate the form that competition takes to ensure the long-term development of the sector. It does not mean eliminating competition, but shaping its terms. The task is to create the right mix of competition and cooperation, a mix that is continuously shifting" (1990, 183).

The long-term sectoral strategic planning has had real effects on the firms' strategies because decision making within the keiretsu firm is relatively unencumbered by short-term stockholder influences. Corporate management is free to pursue the corporation's long-term development without interference from the stockholders unless confronted with a financial crisis. The organizational structure of the keiretsu firm features low formalization, standardization and centralization which allow more influence over strategic decision-making at all levels of the firm. This has been labeled an "organic" structure (Kagano et al. 1984, 45). The top management (which almost always has been promoted on the basis of interpersonal skills and only after extensive experience in the many functional departments of the firm) establishes the strategic vision of the firm. Control is maintained through value sharing and information while relying upon consensual decision making in which conflict resolution is achieved through broad consultation before action. Middle managers coordinate the decision making across functional departments while implementing improvements developed throughout the firm. Lower management seeks to support the production workers in performing their many tasks and acts as a conduit between workers and higher management. Production workers not only perform the typical production tasks, but also, in conjunction with management, are responsible for planning and organizing production, quality control, maintenance of equipment and initiating suggestions for process and product improvements. It has been argued that such an organic structure is superior in terms of adaptability to environmental change (Burns and Stalker 1961; Lawrence and Lorsch 1967).

The organic structure of the firm is particularly adept at incremental change (rather than radical change) and, thus, the crucial ability of continuous improvement has been highly developed within Japan's CIFs.[3] Within the Japanese CIF, "gradual, constant, and incremental improvement" emerges as a normal part of day-to-day operations (Imai 1990, 45). Incremental improvements are small changes in the day-to-day process of work; improvements that, for the most part, would not likely be thought of in a research laboratory. Such improvements, taken individually, often have a minuscule impact on the production process or product, and would often go unnoticed by an outsider. However, over time and in accumulation the incremental improvements alter the system to such an extent that the operation as presently viewed would be considered an innovative breakthrough when compared with a "snapshot" of the operation several years earlier. Matsumoto provides some examples of such an evolutionary process. The "single-step system,"[4] in which the changeover time for the die on a press was reduced from several hours to only a few minutes, was achieved with "no particular epoch-making technological innovation. It [was] simply the result of the accumulation of innumerable small improvements made within the factory,..." (1982, 43). Similarly, he recounts the development of the "one touch joint" system, and the "parallel changeover system" (1982, 43). Just-in-time (JIT) manufacturing is another good example of a breakthrough innovation based on incremental improvements. At Toyota, for example, JIT was the result of an accumulation of incremental improvements which steadily reduced the need for inventories and opened up possibilities for engineers to reorganize machine sequencing to make large reductions in inventories. Those changes both supported and benefited from management efforts to revise relationships with both suppliers and product outlets in a fashion now famous as JIT.

Labor Reforms. Labor reforms initiated during the occupation period have also contributed to the emergence of the CIF in Japan. SCAP imposed the Trade Union Law of 1945 guaranteeing workers the right to organize, bargain collectively, and to strike. This was followed, in 1946, by a law establishing a mechanism for settling labor disputes, and, in 1947, laws were passed to establish protective labor standards, provide a free employment service, and to eliminate feudal labor practices. Union membership rose quickly from less than one-half million to over 6.5 million in 34,000 labor unions during 1945-48 (Kunio 1986, 126-7). Many of the labor unions were, however, dominated by radical and militant leaders, and bitter labor disputes erupted in almost every industry. Strikes were frequent during the latter half of the 1940 and first half of the 1950s. By the late 1950s, changes were becoming apparent in labor-management relations. As the union leadership became more militant, it had moved further away from the beliefs of the rank-and-file workers. This opened up opportunities for the creation of alternative unions that often became more popular with the rank-and-file workers. The leaders of the new unions tended to seek better relations with management and, in return, management often helped them with their organizing activities (Shimada 1994, 262-3). National labor organizations in many industries were subsequently reorganized, effectively strengthening the enterprise unions and providing an arena for intensive interaction between management and labor.

As a result, since the mid-1950s a number of innovative labor-management institutions have become an integral part of the Japanese CIF. The most notable are the joint consultation system, the dual role of first-line supervisors, and quality control circles. The joint consultation system is an important vehicle for communication between management and labor on critical issues affecting the corporation and its workers. Joint consultation is used to discuss a wide spectrum of issues, ranging from broad topics like investment and production plans to specific issues such as the revision of housing allowances. The first-line supervisors also play an important role in communications between management and workers. First-line supervisors have dual roles in Japanese firms; they are the lowest-level manager and the highest-level worker. They are almost always promoted only after years of service and broad experience within the ranks of the production workers. They are very knowledgeable about production operations and the business organization as a whole. Consequently, they play a crucial role in transferring information within the firm. As leaders of production teams they share their knowledge with other production workers and as low-level managers they are a key link between the workers and management. The now well-known quality control circle was designed to bring the production workers into the continuous improvement process. Quality control circle activities contribute to improvements by finding ways the team members can better work together, as well as directly improving the product and production techniques. Quality control circles thus also facilitate communication and information sharing among the workers and between workers and management.

The widespread adoption of the new CIF's institutions and its labor-management institutions have had a significant impact on the way Japanese manufacturers view the employment of human resources. Within the context of the Japanese CIF, human resources are viewed as the most important factor in contributing to the long-term success of the firm. A substantial proportion of the firm's employees is hired upon graduation from high school, technical school or college and anticipates lifetime employment (i.e., becoming "permanent" employees). The firms recruit prospective permanent employees largely without regard to their specific job skills, focusing instead on trying to hire those graduates who they believe have demonstrated a willingness and ability to learn. The essence of the CIF's view of labor is that the firm is making a commitment to a long-term investment in human capital to ensure the on-going success of the firm. Lifetime employment has made it possible for Japan's manufacturers to assume a significant proportion of the cost of investing in the human capital of its employee's. Investment in human capital formation takes the form of both off-the-job training (Off-JT) and on-the-job training. Off-JT is generally characterized by group learning and is usually presented as a series of lectures.

Investment in human capital by Japanese CIFs is, however, most often associated with on-the-job training (OJT). It can take the form of cross-department lectures, education and training within quality circles, one-on-one training of junior workers by senior workers, and job rotation. Cross-department lectures take place frequently within the CIF, especially when a new product or production process is introduced. Dore and Sako state, "a taken-for-granted part of the production engineering involved is for the engineers and work study experts to hold formal sessions to explain the changes, and to instruct - as hands-on as necessary - those who have to do something new just how that new thing is to be done (1989, 90)." This type of OJT extends beyond the CIF's employees. When changes are made in the product or production process, engineers are also dispatched to the suppliers to make sure that their workers are trained in the changes they will need to employ on sub-assemblies. The training and education process, what Okuda calls sociological engineering, is not just a one-way transfer of information. In addition to the engineers providing the production workers with technical information, "engineers must try to encourage and help operators to develop their own ideas and present them in the form of specific suggestions (Okuda 1983, 26)." In this manner the production workers not only learn of the changes which have been developed elsewhere within the firm, but also provide feedback to further improve the product and production processes.

OJT is also a central function of quality circles. All quality circle groups study a standard set of analytical techniques, including simple operational research methods of defect analysis, tree diagrams, Pareto cumulations, statistical process control, and so forth, as a basis for further training and education. Beyond this basic education and training, a group will chose a problem area to work on and it, or a sub-group, will set about to study some aspect that has a potential bearing on solving problems or otherwise improving products, processes, or systems.

One-on-one training between a supervisor and his subordinate and between seasoned and junior level employees is another important source of training and education within the Japanese CIF. The role of a supervisor as a trainer is considered a long-term responsibility and one that ultimately counts very heavily in the evaluations of the managers. Similarly, it is expected that all workers with seniority will pass on to their juniors what they have learned during their years of experience. Since all permanent workers are multi-skilled, sharing knowledge and skills with other workers does not represent a threat to one's job.

Multi-skilling of workers is achieved through what is probably the most extensive form of OJT within the Japanese CIF, that of job rotation. Job rotation is practiced not only with managers and technicians, but also is utilized extensively with production workers. Possibly the greatest impact of job rotation is that it provides all workers access to a general knowledge base of the production processes and an understanding of how the various tasks in the production process are interrelated. This enables the workers to be more effective when offering suggestions for improving the production process and the product. Having an understanding of the various production tasks and how they are performed provides the workers with important information they can use in the quality circles to solve production problems. As a result they are less likely to make suggestions for changing one task that will be problematic for other related tasks in the production process. Job rotation also facilitates the knowledge sharing needed for the important function of horizontal communication within the firm. Job rotation familiarizes workers with the various jobs and enhances their ability to process and communicate information which is necessary for all workers to participate effectively in continuous improvement process (Aoki 1990, 11).

In an analysis of Japanese firms' investment in human capital and the relationship to lifetime employment and seniority wage and promotion systems, Imaoka says that the firm and the trainee share the costs and benefits of OJT investment. "In these systems, the trainee bears part of the initial OJT cost but is afterwards guaranteed its benefits. Those benefits give the employee an incentive to stay with the company that trains him. In this way the corporation's employment structure gains its cohesiveness and rationale. A large part of the skills is most often passed on by employees embued with the corporation's work customs, thus making skills uniform throughout the corporation" (1989, 417-18). The employer bears the expense of the resources allocated to the training and education of its workers, while the employees share the cost by foregoing the possibility of getting an immediate wage increase as compensation for improving their skills. The benefits accrue to the firm over time in the form of product and production process improvements that flow from a permanent labor force that is highly trained and educated. As the firm prospers the employees also benefit, in the long run, as their wage and salaries increase with their seniority, and, as the firm grows, more management positions are created, increasing the probability for promotions. The costs of training and education to the Japanese CIF are consequently held down in the short run, while the long-term benefits are secured for the firm which maintains a permanent labor force that uses its skills to advance the firm. The result is a high level of investment in human capital by the CIF.

Human resources, therefore, are viewed by the Japanese CIF as a capital investment and are the key to the long-term success of the firm. The Japanese CIF has built upon this tradition and focus with its emphasis on incorporating all of its employees (production workers, engineers, and managers) into the continuous improvement process. The current organizational and managerial characteristics of the Japanese CIF create incentives for extensive education and training of its labor force, generally beyond what is feasible for its foreign competitors. To maintain this competitive advantage, Japanese manufacturers must be able to continue making significant investments in the education and training of their labor force.

WESTERN THEORY OF THE MARKET

Neoclassical Theory of the Market

The predominant view of the market within Western economic theory is the neoclassical model of perfect competition. Perfectly competitive product markets are said to be characterized by large numbers of buyers and sellers, a homogeneous product, diffused market power (meaning the individual firm has no control over price), and freedom of entry and exit. The analysis indicates that firms in perfectly competitive markets will produce at the lowest average cost of production, given the available technology, and the price the consumer pays is equal to the marginal cost and the minimum of average total cost. The firm can earn only a normal profit in the long run. By assuming no externalities and perfect information on the part of all participants, the model theoretically proves that perfectly competitive markets are efficient. That is, the marginal social cost of production is equal to the marginal social benefit.

Similar analyses of other product market structures (monopolistic competition, oligopoly, and monopoly) conclude that various degrees of inefficiency will be present in all other markets. Such analyses have long provided the basis for a free market-oriented policy in many Western countries that favors limited government regulation, particularly with respect to the entry of firms into markets. It has also provided the basis for anti-trust policies that either promote competition or prevent monopolies from forming, or attempt to do both.

Contestable Markets Theory. A more recent addition to Western economic theory of the market, known as contestable markets theory, extends the theoretical benefits of limited government regulation to all other market structures. Contestable markets theory focuses on the strategic interactions between incumbent firms and potential entrants. A perfectly contestable market has two properties. "First, the potential entrants can, without restrictions, serve the same market demands and use the same productive techniques as those available to the incumbent firms. ...Second, the potential entrants evaluate the profitability of entry at the incumbents' pre-entry prices. That is, although the potential entrants recognize that an expansion of industry outputs leads to lower prices - in accord with the market demand curves - the entrants nevertheless assume that if they undercut incumbents' prices they can sell as much of the corresponding good as the quantity demanded by the market at their own prices." (Baumol et al. 1988, 5) The first property implies that the entry process is entirely reversible without cost to the potential entrant. That is, upon exiting from a market the entrant can recover any and all outlays that were necessary to enter the market. Hence, the role of sunk costs plays a critical role in determining whether a market is contestable. When entry requires substantial costs that are not recoverable (i.e., sunk costs), entry is not completely reversible and, thus, the market is not perfectly contestable. If, on the other hand, the set of production techniques does not require sunk costs, entry is completely reversible and the market is perfectly contestable. If minimal sunk costs are required then the market is almost perfectly, or imperfectly, contestable.

When contestable markets exist, it is argued that a sustainable industry configuration (a price vector and a set of output vectors, one for each of the firms in the industry) has the following properties: "First, the quantities demanded by the market at the prices in question must equal the sum of the outputs of all the firms in the configuration. Second, the prices must yield to each active firm revenues that are no less than the cost of producing its outputs. And, last, there must be no opportunities for entry that appear profitable to potential entrants who regard the prices of the incumbent firms as fixed" (Baumol et al. 1988,. 5). The properties of perfectly contestable markets and sustainable industry configurations are achieved only in equilibrium. A perfectly contestable market with an unsustainable industry configuration will invite entry that will alter the industry configuration. "Thus, the study of contestable industry configurations is the study of the control that free entry in contestable markets exercises over incumbent firms in equilibrium" (Baumol et al. 1988, 6).

Although perfectly contestable markets have characteristics that are distinct from those of the neoclassical model of the perfectly competitive market, the price-output equilibria of the two models are consistent. In the perfectly competitive market model, potential entrants behave in the same manner as potential entrants in perfectly contestable markets. They both evaluate the profitability of entry on the basis of the incumbent firm's price at pre-entry. Baumol et al. demonstrate that the competitive equilibria "are the only configurations sustainable in the limit as the minimum efficient scale of production shrinks relative to the quantities demanded by the market" (1988, 6). That is, a perfectly contestable market with a large number of firms will result in the same equilibrium as that of a perfectly competitive market, with all of the concomitant implications for an efficient allocation of resources. At the other end of the market spectrum, in the case of the natural monopoly, Baumol et al. show that a single firm can be a sustainable industry configuration as well. To achieve sustainability under conditions of perfect contestability, a natural monopoly can only earn a normal profit and, therefore, must also operate in an efficient manner. Thus, contrary to the neoclassical theory of pure monopoly, contestable markets theory argues that when a market is perfectly contestable a Ramsey-optimal price is the only guaranteed sustainable condition and is thus a necessary condition for equilibrium even in a single-firm market.

In explaining the case of a monopolist, Baumol et al. argue that markets in which the set of production techniques dictate that the size of the incumbent firm's output be large relative to the market demand (i.e., allowing only a small number of incumbent firms) and where the entry process is completely reversible, potential entrants will have incentives to enter the market if the incumbent firm's price is above average cost. An entrant firm does not have to fear price changes by the incumbent because the entrant can exit without cost. The entrant, thus, would earn economic profits at the pre-entry price of the incumbent and then exit if the incumbent reacts in such a manner as to preclude further profit for the entrant. Incumbent firms in contestable markets, therefore, can best protect themselves from successful entry of rivals by setting sustainable prices. Under such conditions, potential entrants cannot expect to earn an economic profit and, consequently, have little incentive to enter the market.

Sustainable prices are not, however, an absolute guarantee against entry. Baumol et al argue that a financially stronger potential entrant that is determined to enter a market may do so successfully even if the incumbent firm's price is sustainable. Their reasoning is that while the entrant is in operation the incumbent firm(s) must lose sales to the entrant. At a lower sales (output) level the incumbent firms' costs will exceed the sustainable price and loses will be incurred. The outcome would depend upon the willingness of the rivals to incur loses in the present in order to secure the market in the future.

What policy implications does Baumol et al derive from contestable markets theory? They conclude that "if an industry is structurally contestable and is behaving accordingly, and if, in addition, it has sustainable configurations available to it, and if none of these involve any significant welfare problems, then that industry is best left to its own devices with no government interference, *even if it is composed of a very small number of firms*" (1988, 466, original italics). What if a market is not behaving in a way that is to be expected in a contestable market? They recommend "the most promising course for public policy is identification and removal of any artificial obstacles to contestability (such as governmental inhibition of entry or other such constraints imposed by public policy, as well as other barriers) and the prevention, so far as is possible, of predatory acts or deterring threat by incumbents. If the industry is not structurally contestable, it may be possible to seek measures that decrease 'natural' entry barriers to some degree" (1988, 466). The removal or diminution of barriers extends to international trade, as well as domestic markets. Only where markets are not contestable or where no contestable solutions are available is direct regulation or attention by antitrust authorities likely to be justified in their view.

Thus, both neoclassical theory and contestable markets theory indicate that a policy of free entry into markets will result in an efficient allocation of resources. Put another way, government regulations that restrict entry into the market imply an inefficient allocation of resources and, generally, should be removed.

COMPETITION IN JAPAN'S MANUFACTURING SECTOR

Despite the brief attempt to reverse the trend in industrial concentration during the occupation period, Japan's markets for manufactured goods have remained dominated by a relatively small number of large CIFs operating within the keiretsu structure. Nevertheless, the vast majority of all Japanese firms have less than 100 employees (i.e., small firms). But in manufacturing almost two-thirds of the small firms are primarily sub-contractors for the large firms. In some industries, including machinery, motor vehicles, clothing and construction, subcontractors produce as much as 70-90 percent of the total value added (Gardner 1988, 195). A relatively small number of large firms, with more than 500 employees, therefore, either directly or indirectly have dominated many of Japan's markets for manufactured goods. In addition, foreign manufacturers have been able to captured only a relatively small share of Japan's markets for manufactured goods.[5]

Contestable Markets in Japan?

It is obvious that such market structures are inconsistent with the perfectly competitive market model. But can they be said to be consistent with the characteristics of perfectly (or imperfectly) contestable markets? As indicated in the above discussion of contestable markets theory, the number of firms in an industry neither determines whether a market is contestable nor whether it is efficient. Contestability is determined by the relationship between the incumbent firm(s) and potential entrant(s). Certainly during the pre-1980s, under the policy of direct regulation, Japan's markets obviously were not contestable. But, in the post-1980s with the shift toward deregulation, the potential for creating contestable markets has been opened up.

As stated in the previous section, a key property of contestable markets is that potential entrants can use the same productive techniques as those available to the incumbent firms. To the extent that Japan's manufacturing sector is dominated by CIFs that emphasize internal incremental technological change, it is practically impossible that such could be the case. The CIF's production techniques are constantly being changed (improved) through an internal process that is unique to each firm, and, therefore, are not readily available to potential entrants. One might reasonably suppose that another CIF could independently develop production techniques that would be sufficiently similar to an incumbent firm's production technique that would largely, if not precisely, fulfill this property. Thus, in the absence of any other barriers, the closer the potential entrant (as a CIF) could come to matching the unique production techniques of the incumbent firm(s), the closer the industry would be to a perfectly contestable market.

Contestability, however, would be particularly difficult to achieve for firms that are not characterized by the continuous improvement features. For example, mass production (MP) firms acquire technological improvements from external sources, that is from R & D units of the firm, from suppliers of capital equipment, or some other source outside the firm's production units. For a MP firm to be able to use the same production techniques as that of an incumbent Japanese CIF would probably necessitate that the MP firm first convert itself into a CIF.[6] But this is something that cannot be done independently because Japan's CIFs production techniques are developed within and supported by a network of firms (i.e., the keiretsu). A potential entrant featuring mass production characteristics would have the following options: join an existing network of firms, combine with other CIFs to form a new network of firms, or combine with other MP firms to form a new network of firms that want to convert to CIFs. It would be very difficult (if not impossible) for a MP firm evolving into a CIF to join an existing network of firms in Japan, since they are rather closed networks. Similarly it would be very hard for such a firm to combine with other Japanese CIFs to form a new network of firms, because existing CIFs, by definition, already belong to a network of firms. It might be possible, in principle, for a MP firm to find a number of CIFs that are members of existing networks that are dissatisfied with them and convince the CIFs to break away and create a new network of firms. A third, and more likely, scenario may be for a MP firm to join with other MP firms making the conversion to CIFs to form a network of firms.[7]

Regardless of the route taken by a MP firm to convert to a CIF in order to acquire the technological capabilities of the incumbent CIF(s), there would be organizational costs incurred in the process. The resources necessary to join or organize a network would not be recoverable in the event the firm exited from the market and did not find the organizational network useful in other markets. That is, there would be some sunk costs associated with entering the market, unless the firm could utilize the organizational network in some alternative market.

In addition to the costs associated with joining or organizing a network, a MP firm evolving to a CIF would additionally incur costs of training their labor force in the production techniques employed by incumbent firm(s) and in the tools of continuous improvement.[8] Because the resulting human capital would be encapsulated within the workers and could not be separated from them, the firm would not own the resulting capital. It would not, therefore, have the capability of selling the capital in the event it chose to exit from the market.[9] Once again, how the firm exits the market would determine the extent of the sunk costs. If the firm could shift the human capital resources to another market that took advantage of the continuos improvement tools and the production techniques, the sunk costs could be avoided. On the other hand, to the extent that neither the tools nor the techniques could be utilized in alternative markets, sunk costs would be incurred. In the case of human capital expenditures, it is almost certain that some of the production techniques would be unique to the particular market and, therefore, would not be transferable to other markets.[10] The less transferable the human capital, the greater the sunk cost to an evolving CIF as a potential entrant and the larger the barrier to entry. The requirement of organizational and human capital investments implies that Japanese markets with incumbent CIFs would be less contestable by MP firms than by CIFs. The greater the sunk costs for organizing a network and training workers, the less perfectly contestable the market would be to a MP firm. Thus, it would seem that because foreign manufacturers are much more often organized as MP firms they would incur greater sunk costs to enter the Japanese manufacturing markets than would domestic firms,[11] which are much more likely to be organized as CIFs.

If the sole purpose of the regulatory restrictions to entry were to protect the Japanese markets from foreign competition, one might argue that the Japanese government decided in the 1980s that the regulatory restrictions have become redundant since the evolution of the CIF. The likelihood that most, if not all, potential foreign entrants would face some organizational and human capital sunk costs would imply that Japan's domestic manufacturing markets would remain incontestable by foreign manufacturers even if government regulations were fully removed. On the other hand, because the domestic CIFs face much lower (and possibly negligible) sunk costs for organization and training,[12] barriers to potential domestic CIF entrants are much lower (and possibly near zero). The complete removal of restrictions on entry would result in (imperfectly) contestable market in Japan's manufacturing markets, even while they remain largely incontestable by foreign firms. Since the organizational and human capital expenditures alone may be sufficient to protect most Japanese manufacturing markets from entry by foreign firms, should we conclude that the Japanese government will continue to move toward free markets in order to attain the potential benefits of an efficient allocation of resources?

Strategic Planning. The consistent use of sectoral strategic planning as a part of Japan's industrial policy suggests that such a conclusion may be unwarranted. As stated above "the purpose of a sector strategy is to negotiate the form that competition takes to ensure the long-term development of the sector" (Best 1990,. 183). The form that competition has taken in Japan's manufacturing markets is a Schumpeterian-type competition that primarily takes place on the floor of the manufacturing facility. The basis for this competition is found in the continuous improvement process. By continually improving the production processes and product quality the Japanese CIF attempts to achieve a competitive advantage over its rivals. Improvements in the production processes reduce costs of production and improve product quality, allowing the Japanese CIF to simultaneously provide a high quality product at a competitive price. Product improvements and flexible manufacturing systems also give the Japanese CIF a dynamic competitive advantage in both shaping consumer preferences and responding to changes in demand. Abernathy, Clark and Kantrow (1983, 73) indicate this in stating, "the primary sources of this advantage are found ... in the Japanese producers' mastery of ... a well-designed strategy based on the shrewd use of manufacturing excellence." Also, Kagano et al. state, "Japanese firms stress finding their 'niche' and strategies of product differentiation in order to gain competitive advantage" (Kagano, et al. 1984, 41).

The key to achieving success in this Schumpeterian competition has been the Japanese CIF's investment in human capital. A significant and on-going investment has been made in human capital by the incumbent firms, much of which is firm specific. In such an environment, it is of utmost importance to the CIF's long-term competitiveness that it maintains its labor force in tact to be able to reap the benefits of its human capital investment. In the absence of barriers to entry, this may be difficult to achieve. Consider a scenario in which the various government regulations that restrict entry into Japan's markets for manufactured goods are suddenly swept away. Contestable markets theory suggests a sustainable industry configuration would result when the incumbent firm(s) earn only a normal profit. Such a case would create no incentive for potential entrants to enter the market. It does not guarantee, however, that new firms will not enter as is pointed out by Baumol et al. (1988). If a potential entrant is determined to enter a market, and has sufficient wealth to sustain itself while suffering loses, it can do so. Given the emphasis that Japanese CIFs place on long-term growth and since they only slowly withdraw from disappointing markets (Kagano et al. 1984, 40-41), it is highly likely that, in the complete absence of regulatory restrictions, potential Japanese CIF entrants would challenge incumbent CIFs, even in sustainable markets, and would be willing to incur losses if they thought they could secure a share of the market for the future.

The result would be over capacity in Japan's manufacturing markets. As Baumol et al. point out, this would result in losses by the incumbent CIF(s) because they would lose sales as long as the entrant remained in the market. The incumbent CIF(s) would therefore be faced with the possibility of terminating some employees in order to reduce its losses and prolong its survival. An alternative for the incumbent CIF(s) might be to employ the redundant human and other capital resources to enter other markets to avoid realizing its sunk capital costs. But such a scenario would likely create over capacity it those markets and threaten the incumbent CIF(s) there. They could either realize their sunk costs or choose to employ their redundant human and other capital resources in still a third market. In short, free entry into Japan's markets would create the possibility of a domino effect in which the Japanese CIFs' investments in human capital would become a more risky endeavor and thus they would be less likely to make such investments. If Japanese firms shift away from making investments in organizational and human capital, their strategic competitive advantage as CIFs will have been eroded. In the long run the erosion of the their strategic competitive advantage would be detrimental to the firms success and survival. Therefore, the static efficiencies of deregulation may be fully or partly offset, in the long run, by dynamic inefficiencies peculiar to the Japanese CIF.

CONCLUSION

During the post-WWII period Japan's industrial policy has served, to varying degrees, as a barrier to entry into its markets for manufactured goods. Western economic theory suggests that the government regulations have imposed inefficiencies on the Japanese economy and, therefore, should be removed. In addition, the evolution of the CIF in Japan, and the inability of foreign MP firms to readily acquire the technology employed by the incumbent Japanese CIFs, has resulted in another layer of sunk costs and added to the barriers faced by foreign MP firms. The post-1980s trend toward deregulation may appear to suggest that Japanese government sees the evolution of the CIF as an adequate protection for the domestic market and, therefore, will continue to move toward a free market policy. To the extent, however, that sectoral strategic planning continues to be a part of its industrial policy, selective regulation may continue to serve a purpose in Japan. One might conclude that the Japanese government believes that the long-term strategic efficiencies are more important than short-term efficiency in some industries.

Nevertheless, the future implications of the post-1980s trend toward relaxed government regulation on entry into Japan's industrial markets may not be so kind to Japan's CIFs. In those industries where government regulation on entry is relaxed it is likely that increased competition among domestic firms will increase initially. If industries that had not achieved a sustainable configuration prior to relaxation of regulations on entry, increased competition will likely reduce price and increase the variety of goods available. This would move the industry toward a sustainable configuration and increase efficiency. In other markets that had achieved a sustainable configuration prior to relaxation of the regulations on entry the increased competition may be a harbinger of significant institutional changes for Japan's CIFs. Entry of firms into a market with a sustainable configuration will likely cause the incumbent CIFs to shift redundant resources to other markets, ad infinitum. Eventually, Japan's CIFs will have to confront the choice of maintaining its labor force in tact, or continuing to pursue aggressively long-term growth. If they continue to pursue long-term growth by aggressively entering markets with a sustainable configuration, then the CIFs, in the aggregate, will be faced with the necessity of reducing their redundant resources to cut their losses and survive for the long term. If, on the other hand, they choose to forego aggressive pursuit of long-term growth, over capacity will be less of an issue, but CIFs will be faced with making internal adjustments necessitated by a slow-growth environment.

Either choice, however, will create some potential difficulties and likely bring about institutional changes for Japan's CIFs. Pursuit of long-term growth through aggressive entry into sustainable markets will likely cause the CIFs to shift away from the lifetime employment and the seniority wage system. This may undermine the continuous improvement process. If workers can no longer count on employment security they may become less willing to obtain the firm-specific training and skills that are the basis of the continuous improvement process. And when they do obtain new skills they may be more inclined to demand immediate pay increases as compensation for their increased productivity. If CIFs immediately have to pay for higher skill levels, the cost of investment in human capital increases and the CIFs would likely decrease the amount of training they provide to their labor force. Either way the continuous improvement process will be harmed by these changes.

Alternatively, Japan's CIFs could curtail their aggressive pursuit of long-term growth. This would necessitate restraint in entering markets that have achieved a sustainable configuration and would mean slower growth for the firm. Slower growth, however, is also likely to be detrimental to the continuous improvement process. The motivation for participation in the continuous improvement process comes, at least in part, from the expectation of future growth and the concomitant employment security and promotion. In the absence of high rates of growth, long-term employment is less secure (especially if productivity is increased) and promotions will be less available as fewer new positions will be created in the slow growth environment. Here again, curtailing aggressive pursuit of long-term growth has adverse consequences for the CIFs crucial ability of continuous improvement.

What about the potential effects of relaxed regulation on entry for foreign firms? Given that government regulation is selective and indirect, it may not greatly expand the opportunities for foreign firms. If the government regulators use selective rules that put foreign firms at a disadvantage, entry may remain limited. If government regulation is relaxed in a manner that treats domestic and foreign in a like manner, there still may not be much immediate opportunity for entry by foreign firms. As long as the CIFs maintain their continuous improvement capabilities and foreign firms lack those capabilities, significant barriers to entry in the form of sunk costs for reorganization will exist for most foreign firms. Over time, however, if the relaxed regulatory environment causes Japanese firms to reduce their reliance on the continuous improvement processes (as indicated above), the opportunities for foreign firms may increase as they would be better able to acquire the technology used by the incumbent firms in the market without incurring the sunk costs of restructuring their firms to become CIFs.

The recent relaxation of direct government regulation of entry into Japan's industrial markets implies that Japan's markets have already become more contestable by domestic CIFs. This increased contestability implies that significant future institutional changes are likely for Japan's CIFs. An indirect consequence of the future institutional changes in Japan's CIFs is that foreign firms may find it less costly to enter Japan's markets for manufactured goods and, thus, make Japan's markets more contestable by foreign non-CIF firms. If these long-run changes transpire, it is likely that the Japanese firms and regulators will once again find it in their mutual interests to move back toward greater regulation of market entry. To conclude, therefore, that the recent changes in Japan's industrial policy represents a movement toward convergence with U.S. or Western policy and a free market economy would seem to be wishful thinking.

ENDNOTES

1 It should be pointed out that this is not Schaede's view, but rather he is paraphrasing the views of others expressed in several articles that appeared in *The Economist*, *Business Week*, and *The Wall Street Journal*.

2 Undesired activities were ruled out a priori.

3 We emphasize that the "continuous improvement firm" is a Weberian ideal type whose characteristics may not be fully reflected in any single firm (possibly the closest fit would be Toyota Motor Company). Certainly, not all firms in Japan are of the continuous improvement type. Many firms deviate considerably from the continuous improvement mode. Significantly, however, many of the firms (and their keiretsu member firms) that are identified as "world class" performers in the automobile and electronics industries have adopted continuous improvement as their dominant strategy.

4 The "single-step system," referred to by Matsumoto, is more generally known as the "single-minute exchange of dies," or SMED.

5 While imports of manufactured goods into Japan have increased in recent years, this has often been the result of negotiated trade, such as the agreement between the U.S. and Japan wherein the U.S. computer chip manufacturers' sales were to equal a minimum of 20 percent of the Japanese market.

6 For the purpose of this analysis, we are ignoring the issues related to the difficulties a MP firm might have in converting to a CIF. For a discussion on this topic see Cole and Mogab (1995, chapter 10).

7 Another alternative might be to combine the second and third scenarios, whereby one or more dissatisfied CIFs combine with MP firms converting to CIFs to form a network.

8 There are a number of tools employed by CIFs. Some of the most common include: statistical process control, Pareto diagrams, histograms, scatter diagrams, and cause-and-effect diagrams.

9 In some cases potential entrants may also need to make expenditures for physical capital, but unlike human capital, physical capital is owned by the firm and can be sold upon exiting the market and, therefore, has a greater likelihood of at least partial recovery. In addition, as emphasized above, human capital plays a crucial role in the operation of a CIF and represents the most important investment on the part of the firm. In addition, since most foreign MP firms do not have a policy of lifetime employment, they would find it difficult to invest in human capital to the same extent as Japan's CIFs.

10 The human capital costs related to duplicating the production technique of the incumbent firm(s) may apply equally well to existing CIFs as potential entrants.

11 Those Japanese firms that are organized as MP firms would face the same sunk costs as foreign MP firms as potential entrants into a market in which the incumbent firms are CIFs.

12 They face no marginal costs for organizing a network or training in the use of continuous improvement tools, but may have additional costs of adopting a similar production technique used by the incumbent firm(s).

REFERENCES

Abernathy, William J., K. B. Clark and Alan M. Kantrow. *Industrial Renaissance: Producing a Competitive Future for America.* New York: Basic Books, 1983

Aoki, Masahiko. "Toward an Economic Model of the Japanese Firm." *Journal of Economic Literature* 28(March 1990): 1-27.

Baumol, W. J., J. C Panzar and R. D. Willig. *Contestable Markets and the Theory of Industry Structure*, Revised edition. San Diego: Harcourt Brace Jovanovich, Publishers, 1988.

Best, Michael H.. *The New Competition: Institutions of Industrial Restructuring.* Cambridge, UK: Polity Press, 1990.

Burns, T. and G. M. Stalker. *The Management of Innovation*, London: Tavistock, 1961.

Cole, W. E. and J. W. Mogab. *The Economics of Total Quality Management: Clashing paradigms in global competition.* Cambridge, MA: Blackwell Publishers, 1995.

Dore, R. P. and Mari Sako. *How the Japanese Learn to Work.* London: Routledge, 1989.

Gardner, H. Stephen. *Comparative Economic Systems.* Chicago: The Dryden Press, 1988.

Imai, Masaaki. "The Kaizen Wave Circles the Globe." *Tokyo Business Today* (May 1990): 44-48.

Imaoka, Hideki. "Japanese Corporate Employment and Personnel Systems and Their Transfer to Japanese Affiliates in Asia." *The Developing Economies* 27:4(December 1989): 407-26.

Kagano, T., I. Nonaka, A. Okumura, Y. Sakakibara, and A. Sakashita. "Mechanistic vs. Organic Management Systems: Comparative Study of Adaptive Patterns of American and Japanese Firms." In *The Anatomy of Japanese Business*, K. Sato and Y. Hoshima, eds. London: Croom Helin, 1984.

Kunio, Yoshihara. *Japanese Economic Development: A Short Introduction*, Second edition. Tokyo: Oxford University Press, 1986.

Lawrence, P. R. and J. J Lorsch. *Organization and Environment: Managing Differentiation and Integration.* Boston: Harvard Business School, Division of Research, 1967.

Matsumoto, Koji. "The Secret of Japanese Management Resulting in High Productivity, Part II." *Journal of Japanese Trade & Industry*: (March 1982): 40-45.

Okuda, K. "The Role of Engineers in Japanese Industry and Education." *Journal of Japanese Trade & Industry* 2:5 (September/October 1983): 23-26 .

Sakai, K. "The Feudal World of Japanese Manufacturing." *Harvard Business Review*: (November/December 1990): 38-40.

Schaede, Ulrike. "Change and Continuity in Japanese Regulation." *Berkeley Roundtable on the International Economy Working Paper* # 66 (1994).

Shimada, Haruo. "The Perceptions and the Reality of Japanese Industrial Relations." In *Comparative Economic Systems: Models and Cases*, 7th edition, Morris Bornstein, ed. Burr Ridge, IL: Richard D. Irwin, Inc., 1994.

3

THE ROLE OF ORGANIZATION IN SHAPING THE TECHNOLOGICAL CAPABILITIES OF FIRMS: AN EXAMINATION OF THE CONTINUOUS IMPROVEMENT FIRM[1]

Stella S. Schramm
Vice President, Economic Development
Chamber of Commerce, Oak Ridge, Tennessee

INTRODUCTION

In some market segments, technology is changing at a dizzying pace. With this development, the importance of technological capability has become an inescapable reality of the marketplace. A highly visible group of market entrants in this milieu have been Japanese industrial giants called "continuous improvement firms" (CIFs) because of their continuous improvement of products and production practices (Imai 1986). These firms have made dramatic inroads in their markets and have redefined product and production standards in industries that were previously considered to hold little opportunity for technological innovation. In this environment, the ability to develop and employ technology has achieved heightened importance in both international competition and in economic thought.

As many firms strive to copy CIF practices, economists are confronted with questions of how to determine key factors shaping the technological capabilities of firms. For development economists, these questions are new faces on the old issues of technological access and capability: namely, what institutional factors shape the success and transfer of CIF methods? The author describes elsewhere (Schramm 1994) a system of technology development employed by CIFs that is characterized by a concurrent functional and temporal, rather than a sequential, approach to innovation. This approach is demonstrated to have a dramatic impact on competitive ability and to depend upon a specific means of industrial organization. The purpose of this essay is to articulate this organization and to explore its effect on technological capability. In this way, the essay seeks to make a small contribution to understanding the key factors shaping the CIF's ability to innovate continuously and to relating these findings to economic analysis of technology and innovation. The approach used here is, first, to assess the internal dynamics shaping the technological capability of firms. Second, these findings are used to compare the organizational paradigms of CIFs with the conventional Scientific Management/Mass Production (SM/MP) firms frequently assumed in economic theory.

ECONOMICS MEETS THE CIF

Before exploring the factors shaping the technological capabilities of CIFs and other firms, it is useful to consider how technological capability is currently addressed in economic analysis. Technology is widely assumed to evolve externally to the directly productive activities of the firm. For the individual firm, the opportunity to acquire new technology occurs only when the firm invests in new capital equipment.[2] Economic models of the firm have, therefore, treated technology as an exogenous factor likely to change only over the long run as firms are able to replace capital equipment. Based on this reasoning, models of firm competition assume that each firm is using the best practices for its industry. Thus, differences in technological capability are not generally considered, except as long run investment and market entry timing decisions. From this perspective, apparent differences in technological capability can be interpreted as differences in technology strategies. Should a firm fail to be competitive due to genuinely inferior abilities, its demise can be viewed as a boon to the efficiency of the marketplace. Since technological capability is assumed to be equally attainable by firms of any national origin, there is no need for concern about the loss of uncompetitive industry, as long as the market is allowed to function freely. Only in development economics has access to technology and disparities in technological capability been the subject of much concern.

The accepted economic concepts for explaining changes in a firm's productive and technological capabilities are learning (as described by Arrow's depiction of the learning curve) and cost/benefit (C/B) analysis. According to Arrow's concept of learning-by-doing, a firm may experience a gradual increase in productivity following a technological investment. This gain is attributable to the firm learning to use newly installed capital technology. Other than this initial learning effect, a firm's technological and productive capability is defined by the amount and timing of its capital investment. Based on this concept, there are limits to the improvement a firm can make with any given level of capital investment. Beyond initial learning gains, a firm will need to make new capital investments to make further performance gains.

In economic thought, the mechanism for evaluating opportunities for technological change within the firm is cost/benefit analysis. According to this thinking, if opportunities for internal improvements exist, they will surface through the firm's own efforts to maximize economic gains. The use or nonuse of any given technology or the initiation of any change is based on the firm's evaluation of the costs and benefits associated with the change. It is generally assumed that any firm can acquire a technology if it is willing to pay the price for it. The decision facing the firm, therefore, is whether the technology will generate more income than its cost. In this light, the initiation of improvements in any firm is simply a matter of the firm's investment decisions. Likewise, the use or nonuse of continuous improvement (CI) techniques, and the resulting stream of improvements to a firm's performance, can be judged as a choice based on cost/benefit analysis.

In the case of continuous improvement methods, however, there are complicating factors in relying solely on learning and cost/benefit concepts to convey the options (and costs) available to firms. First, the limits imposed by learning theory on the potential for innovation within the firm (without substantial investment in new capital equipment) run contrary to observations of gains made by CIFs. Watanabe (1991) cites improvements numbering in the millions for a single firm in the space of a year. Second, it has been suggested by some writers (most notably Best 1990) that CI represents a sea change in industrial organization tantamount to a third industrial revolution. Thus, the abilities and decision-making properties of firms assumed in economic theory are based upon a specific mode of industrial organization that is not shared by CIFs.

Based on past patterns of industrial organization, economic analysis has relied on investment (generally in externally developed technology) to represent the opportunity for innovation within firms. In cost/benefit analysis, investment prospects are centrally evaluated and decided upon as competing uses for scarce resources. In this system, projects for which funding does not present a significant impediment are unlikely to receive formal review or attention.

The cost/benefit approach to seeking improvement opportunities contrasts sharply with observations of CIF innovative behavior. The millions of improvements cited by Watanabe (1991) in one CIF are described as undocumented and undocumentable. Central pooling and competition for projects would not appear to be a feasible means for evaluating such an endless stream of improvements. Nor would a C/B system address the equally important issue of how the firm should go about improving its technological and productive capabilities. Questions of greater technological and strategic importance to the firm remain unaddressed; such as how to generate good projects, how to implement and monitor projects effectively, or how to give direction to the firm's improvement efforts.

A C/B system for addressing the opportunity for innovation within a firm remains only one step removed from an ad hoc approach to making improvements. It does offer a means for comparing competing projects on the basis of a project's identifiable and quantifiable cash streams, but it ignores many aspects of a project's importance to the firm and presents a formidable burden (in both real time and labor hours) that does not contribute to the value of the project. On the level at which project proposals are generated, the effort often is of an ad hoc nature, increasing the time, effort and disruption required to submit a proposal for consideration.

A further complicating factor in using conventional economic assumptions to analyze the CI phenomenon is the apparent difficulties in the diffusion of CI technologies. As with past revolutions in industrial organization, would-be copiers of the new methods find themselves at loggerheads with past approaches and mindsets. In this case, cultural beliefs appear to be significant in shaping even an industrial society's ability to adopt technology.[3] What is apparent in the market today is that even if firms see the value of a CI organization, their ability to comprehend fully and copy the technology may be significantly impeded or delayed, forcing many firms to compete without the benefit of the technology they would choose to use. The delay in acquiring CI technology may be considerably protracted, as well, from the impediment imposed by their reliance on an inappropriate economic paradigm for understanding and interpreting the actions of CIFs.

The conventional assumptions in economic analysis regarding technological change have been workable in a world in which the rate of technology change was relatively slow and in which the organization of technology development was relatively universal. The relevance of these assumptions to competition today depends largely upon the degree to which market conditions diverge from past assumptions of the organization, frequency and diffusion of technology. In many markets, disparity between recent industry trends and common economic assumptions demonstrate the need to consider explicitly organizational factors in assessing the ability of firms to compete in an environment of rapid technological change.

THE MAKINGS OF A CONTINUOUS IMPROVEMENT FIRM

Determining the key organizational factors shaping the technological capability of CIFs is not a facile task, in part because CI technology has yet to be fully understood. The CIF is often described in terms of a laundry list of production techniques and management characteristics. This has given rise to the term "stylized facts" (empirical regularities) to describe the nature of current research findings on CIFs.[4] This has also led to a debate on whether CIF techniques can be transferred separately or must be a complete "bundle" of CIF techniques.

This essay treats the techniques attributed to CIFs as themselves the product of a system for generating technological improvements. The CIF's technology development system consists of two parts: kaizen and concurrent engineering.[5] This characterization combines the kaizen process described by Imai and Watanabe and the product design process observed by Clark and Fujimoto[6] as parts of an overriding approach to technology generation and use that is a significant departure from the approach to technology assumed in economic models of the firm. The CIF technology development system includes those effected by a technology in the development and adaptation of that technology. This system relies upon simultaneity and joint participation to enhance communication and coordination, and to reduce time requirements. This contrasts sharply with the linear approach assumed in conventional models of technology development, in which specialists work sequentially on the development of a new technology.

The CIF's concurrent approach to technology development and implementation is itself dependent on an organizational structure supporting its operation. Since there is no divine schematic blueprint from which to derive the CIF's approach to industrial organization and the principles by which it functions, we are left to discern its structure and rationale from its observable characteristics and capabilities. In this endeavor, we seek to describe the industrial organization upon which the CIF is based in a generic or generalizable sense able to span a wide range of settings.[7] While the much commented CIF techniques and characteristics reflect this organizational framework, we would still expect these techniques and characteristics to vary over time and place as they are adapted to differing needs and opportunities. We are seeking to delineate the capabilities we would expect to see, no matter the specific context or time period.

For our analysis, we use a deductive reasoning approach. We start by identifying the properties a firm would likely possess in order to be capable of continuous improvement. We then compare these properties to observable characteristics of CIFs and the conventional industrial (SM/MP) organization assumed in economic theory and analysis. Organizational characteristics that appear to positively or negatively impact CI capabilities are assumed to be highly significant to a firm's technological and competitive capability and important to the assessment of CI technology transfer. It is the ability of a firm to improve continuously upon its performance that will determine its ability to keep pace with CIFs and ultimately is the criterion upon which the full transfer of CI technology and capability will be judged.[8]

THE REQUIREMENTS OF INNOVATION

In analyzing the necessary components of an organization capable of CI, we start with the most basic and central capability: what is necessary in order to be able to generate and implement a continuous stream of improvements? It has long been assumed that innovative activity is beyond the reach of the average population or the normal work-a-day environment. The very existence of the CIF with its evidence of routine and widespread innovation has proven this assumption false.

To move beyond the assumption that innovation is outside of the scope of mundane production activities and the day-to-day activities of ordinary people, we ask what truly is necessary to discover and implement new methods in production?[9] Changes can be made to a production process with relative ease. What presents a greater challenge is to make gains in overall firm performance as a result. To actually improve a process means to take an ongoing production process, make a modification of some kind, and have the process function better after modification than before. While on the surface this appears quite obvious, this apparent simplicity masks complexity in execution. Because change is disruptive to routine and can have far reaching impact on many aspects of a production environment, having a net positive impact can be an elusive objective.

A key challenge facing a firm in attempting to undergo continuous change within its production activities is how to keep change from disrupting its operations. Meeting this challenge would imply having a continuous source of ideas for improvements, a means for evaluating and refining these ideas into plans, the resources and skills to implement plans, and a means of monitoring and modifying changes once implemented. The net effect of the changes on the firm's operations depends upon how well these things are done and the effect of these demands on other aspects of the firm's operations. A second challenge of CI is how to insure that the changes made are relevant to the firm's most pressing needs and are significant in their contribution to the firm's performance. This would require having a means of directing the firm's innovative energies without stifling or encumbering change efforts.

When looking within the process of change, we see that change puts great demands on an organizational system. As production methods change, the individuals performing the changed tasks, and possibly those supplying the inputs and using the outputs of those tasks, must be able to change their behaviors in concert with the new methods. Changes in production processes can require new skills and work arrangements, and can bring about changes in related aspects of design, such as in product characteristics and downstream processes. If unanticipated prior to initiating a change, repercussions will be felt once a change is implemented. These repercussions can include quality failures, increased resource use per unit, imbalances in production leading to inventory and handling problems, insufficient operator skill to carry out properly and efficiently the modified production tasks, equipment breakage, and slowdown in output levels. These repercussions are the threat of every process change. In addition to efficiency losses and delays in production, their cost to the firm can include erosion of the firm's customer base if reliability in the myriad of aspects of firm performance[10] is in any way adversely affected. These indirect costs can be of far greater significance to the firm than the direct costs of production. Conversely, any improvement the firm makes in these areas can enhance the firm's customer base.[11]

Thus, fully capturing the benefits of process change requires working out the details to make the necessary parallel changes in other areas. Failures in production performance while embarking upon a modified production process have the potential to greatly disrupt a firm's operations and affect its ability to participate in future competition. When a firm changes its production methods frequently, there is the potential to have a jagged edge of performance (rather than a smooth rising learning curve), with productivity dropping with the implementation of a change, rising as necessary parallel changes and skill development are acquired, then dropping again with the next modification. For this reason, a firm has cause to fear the disruption of change, for which it is not thoroughly prepared. Constant change has the potential to generate chaos.

These experiences with change are not necessarily unavoidable, however. Rectifying these problems entails determining, preparing for, and following through on the needs of the modified production process. Knowledge of how to modify a process to fit smoothly into the activity of the firm can be acquired before implementing a change or as a result of trial and error once the modification is in place and production is under way. The information on how to integrate fully changes can be accumulated at once or over many years. Whether acquired quickly or prior to making process modifications will, however, impact the operations of the firm. The postponement of fully working out the details for the thorough integration of a change into a firm's operations will gain the firm time for discovering and implementing adjustments, but will cost in the suboptimal functioning of the production system.[12] To not make necessary related changes quickly and prior to initial modification will limit the frequency and degree of change a firm can afford.

We can distill these demands on an organization into four basic categories of needs. In order to improve routinely a production process, it is necessary to:

1) know something of the process (both to recognize opportunities and to integrate change),
2) know what change is desired and relevant,
3) be able to implement change effectively, and
4) insure continued smooth operation of the firm.

Changes can be made without satisfying these requirements, but they can not enhance the operations of a firm without intentionally or unintentionally fulfilling these four criteria.

When a firm undergoes change infrequently, the penalty for deficits in these areas may be hidden. In an environment in which there is frequent change, however, the cost in performance failures can be crippling. Mastery of frequent change, therefore, depends upon thorough research, planning, and advance preparation to insure appropriateness to the smooth functioning of the firm. Quick changeover between techniques is also critical to minimizing interruptions to production.

Fulfilling these objectives consistently implies a shared understanding of how to go about seeking, implementing and monitoring process improvements. A systematic approach to improvement would include a means for collecting and interpreting information on existing operations and opportunities for change. Such an approach would rely upon tight discipline in the implementation of plans, as well as careful monitoring of production processes before and after changes are made.

Frequent change also puts great demands on individuals working within the system. To accomplish frequent change without deleterious effects on production timeliness, product quality or costs requires a system in which there is close coordination among all individuals affected by each change. Much changeover to new methods would need to occur off-line in order to minimize deleterious effects. It is particularly important that there be minimal organizational resistance or hurdles to change, and that there be widespread ability across the organization to quickly and routinely collect and use detailed information on existing operations, opportunities for improvement, and the ramifications of change. As changes must be universally acted upon; the agreement, coordination, and dependability of all affected parties would appear to be critical. Undergoing frequent change would also depend upon the participation of those most involved in making and living with changes and an absence of lengthy delays for approvals, resource allocations and paperwork. Organizationally, a system with these capabilities would rely upon clear understanding of what is to be accomplished by all involved; individuals accepting mutual responsibility to get the job done; trust to give authority and resources directly to those involved in the change process; and widely distributed skills for generating and evaluating possible changes, making decisions, and monitoring results.

While these needs in managing change can be met on an ad hoc basis, they can also be approached using a codified approach, outlining specific tools and steps for gathering relevant information, evaluating possible methods and outcomes, and implementing change. With the use of a coherent system for change, skills in using the system can be passed on to every employee and associate of the firm, insuring standardization in approach and providing dependability even in the midst of change.

From this analysis, it appears that detailed scrutiny of relevant information and tight discipline and coordination among workers, if not our conventional understanding of control, are fundamental components of a firm's ability to change its technology frequently without deleterious effects. Further, to be most successful, these characteristics would permeate every aspect and level of an organization. While the specific content of future changes and necessary behaviors can not be dictated a priori, an organization can guide how each change is to be handled through the use of organizational tools. This includes establishing methods for determining what and how changes are to be made, setting the criteria changes must satisfy, and providing the means for clear communication and agreement among all affected parties of what is to be done, by whom and when. The effectiveness of these methods, however, would depend upon clear understanding and adherence by all participants.

These organizational directives stand in stark contrast to recent exhortations that, in order to promote innovation and change, management needs to loosen control and promote more risk taking ("entrepreneurial") behavior by individual workers within the firm. Based on our examination, such an uncontrolled approach would likely create havoc in an organization and greatly injure a company's performance. Rather than reducing organizational control, the challenge presented by technological change is how to support and promote change while assuring the performance standards of the firm. Thus, it is not whether to control the activities of the firm, but how to do so while providing the initiative to continuously improve upon the firm's activities.

Control methods would need to support efforts to innovate throughout the firm. The dilemma of how to promote widespread and CI while carefully safeguarding the firm's performance suggests that controls need to be diffused among the labor force, building competence and authority to institute change in those most actively involved in the firm's productive activities.

In diffusing the ability to evaluate, initiate, and coordinate actions; an organization can be responsive enough to discover and implement useful changes while averting performance problems.

If these capabilities can be accepted as the key characteristics affecting the success with which a firm can generate and implement improvements to its products and processes on a continuous basis, then organizational structures can be evaluated and compared for their effectiveness under conditions of rapid technological change.

CONTRASTING ORGANIZATIONAL APPROACHES

Organization affects how a company and its individual workers are able to behave in the range of circumstances faced by the firm. The method by which activities are coordinated also determines the resilience of those activities to change. This section evaluates how two organizational paradigms, SM/MP and CIF, each handle the task of generating and implementing appropriate change. For our purpose of evaluating the technological capability of each type of organization, we approach each of these organizations as a means to an end: that is, we judge them on the effectiveness of their method of directing, coordinating, and controlling the activities of the firm. We focus especially on the ability of each type of organization to initiate CI without sacrificing its overall performance.

These two approaches to industrial organization base their structure on very different organizing principles. The SM/MP firm is based upon a hierarchical organization and chain of command. This has been the conventional means of industrial organization since the first industrial revolution. As pointed out by Adam Smith in 1776, firms are organized to allow specialization of labor and equipment, in order to increase the speed with which each of the firm's activities can be performed. Since the time of Adam Smith, the specialization of resources has gone further with every evolution of industrial organization. Because, in this approach to organization, workers and machines are often focused upon and skilled in only a minute portion of the firm's operations, the firm depends heavily upon supervisory and planning staff to coordinate and direct the activities of those directly involved in production. In order to have a high degree of specialization, it is also necessary to have sufficient activity to employ fully an army of specialized resources. Since specialization of labor and equipment can reduce the time needed to perform individual tasks, yet requires sufficient volume to be fully utilized, specialization is strongly linked to economies of scale for today's SM/MP firm.

In concert with this organizational rationale, the SM/MP firm bases its structure on specialized work activities. Each division of the organization represents a group of specialists and is devoted to performing tasks related to a particular function. Work is routed through the organization, moving from one specialty group to the next, until every necessary task has been performed. Workers are hired by the SM/MP firm on the basis of specific skills. In this manner, the SM/MP firm pays only for the skills needed to perform a given task. As workers' skills increase, their value to the firm increases as well. Promotion within the firm, likewise, is within a skill group (or functional) area. Worker performance is evaluated on the basis of individual achievement of assigned tasks, rather than on how an individual's actions affect others and the overall performance of the firm. It is not until the top position in a functional area is reached that a worker is likely to be called upon to communicate with and to consider the needs of other functional areas. Even in this case, consideration of others is likely to be less important to performance evaluation than one's own achievements considered in isolation from other activities within the firm.

Decision making authority in the SM/MP firm is based upon a chain of command. Decision making authority over a functional area rests with the lowest supervisory level position to bear responsibility for the area. A higher level decision takes precedence over a lower level one. Input to decision making need only include those higher in the chain of command. If a decision obviously impacts others outside of a functional area, it may be communicated to others, although outside input is generally not required or sought. If activities within one area of the firm adversely affect others, the affected parties may present their case to a higher level of command that encompasses both parties for conflict resolution. Those under a decision-maker's command have little assurance of input to or recourse in protesting decisions.

The coordination of such a functionally segmented system depends upon central planning and control. Central planning and control requires time for information to be collected and analyzed, decisions to be made, and directives to be issued and executed. The burden of centrally researching, planning and coordinating actions is onerous. In this respect, the SM/MP firm experiences the same difficulties pointed out regarding centrally planned economies. Knowledge of central planners is necessarily limited and fails to reflect all the opportunities available to the firm. As in centrally planned economies, this results in lengthy time delays for the firm to change its activities or methods. Because information requirements of central planners must be defined without the planners having direct experience with or observation of the particular circumstances for which they are gathering data and making decisions, communication and information handling is awkward. Problems may take time to become apparent to upper management, while signs may exist on the shop level at a much earlier stage.

The dedication of direct labor and capital resources to specialized and minute parts of the firm's activities, the resulting need for high levels of coordination and central planning, and the reliance on a hierarchical chain of command in decision making makes the SM/MP firm a stable but relatively inflexible structure. Labor and capital can not be easily redirected, and input from those directly involved in production is difficult to obtain and is potentially disruptive to operations.

Because of these characteristics, change in the SM/MP firm is most effectively communicated and coordinated if it is conceived by and emanates from the top of the chain of command. Because each worker has so much of his or her corporate value tied to the specific deployment of the firm's resources, change in the configuration or use of resources is likely to be unduly painful to some and enriching to others. Since communication within the firm is predominantly vertical rather than horizontal, the cross-functional coordination needed to implement and integrate changes is likely to be weak.

These difficulties are inherent to the organizing principles undergirding SM/MP firms. Because the firm's structure and task assignments are tied to the specific methods and activities in use, the SM/MP firm is likely to need to restructure in order to make a significant change in its activities. These difficulties greatly add to the cost and disruption of change to the SM/MP firm. Because of this, the SM/MP firm is likely to consider change infrequently (even to resist change) and to save up small changes in favor of infrequent monumental efforts. In an environment of rapid change, firms using this type of organization are likely to find their system too inflexible to make rapid and frequent change without considerable consternation.

In this sense, the SM/MP firm is comparable to the structure of a building: it is a framework intended for a relatively stable technological environment and is not built to move.

In contrast with this, the CIF bases its organization on the speed and accuracy with which the firm can produce exactly what each customer wants at the time the customer wants it. Toyota president Toyoda first envisioned this system based on the pull concept of supermarket inventory (Monden 1983). This perspective has several consequences for the CIF's approach to organization. In order to exactly meet customer demand, the CIF focuses on anticipating customer wants and being flexible in its methods, products, and production. To be able to do this and sell products at a marketable price requires the CIF to gain economies equal to the economies of scale enjoyed by the SM/MP firm. To do this, the CIF stresses the overall effectiveness of its system of production from product concept to satisfied customer, rather than the speed of production for a single worker or machine. Defining the relevant speed of throughput as encompassing the entire production and distribution process has focused attention on reducing the time necessary to change production from one product to the next. This means that the flexibility of labor and capital and the smooth integration of the various activities of production are crucial for the CIF to produce economically and effectively.

In keeping with this idea, the CIF hires on the basis of generalizable skills and rewards workers for acquiring competence in a broad range of the firm's activities. Unlike within the SM/MP firm, the individual worker is responsible for integrating his or her activities with others. Inclusion in decision making is based on the scope of those who will be involved in implementing or are affected by the consequences of a decision, rather than on a chain of command. Control of the behavior of individuals within the firm is based upon adherence to specific behavior guidelines, rather than adherence to prescribed actions and policing from supervisors. Instead of a performance evaluation criterion based on individual achievement alone, evaluation is tied to following the principles of the firm and the effect of one's actions on the overall long-term success of the firm.

Guidelines for workers provide the mechanism for the CIF to insure coordination and cooperation within the firm and provide the means by which improvements are sought, investigated and implemented. These guidelines, in essence, promote the smooth evolution of the organization by requiring continuous review and examination of operations and by detailing how change is to be carried out. In effect, every individual is given the tools to review situations and make appropriate decisions as a manager would do in the SM/MP organization. A key difference, however, is that an individual seeking to make a change to the activities of the firm must include others who are affected by the decision, but need not get upper level approvals if all affected parties are in agreement. This gives individuals the opportunity to make changes as opportunities occur without disrupting the activities of the firm and without lengthy delay. It also allows changes to be implemented smoothly, as those affected by the change have insured the appropriateness of the change to the success of their activities. In this way, the structure of the CIF functions as a wheel, allowing the organization to progress in its methods and products without requiring restructuring to do so.

Thus, we can view observed CIF characteristics as reflections of a fundamentally different approach to coordination of a firm's activities, in which individual technologies and their associated tasks and skills are less important than the core methods used throughout the firm. This approach has major consequences for the firm's valuation and treatment of labor, approach to technology development, and flexibility in its operations. Because labor and capital are more flexible using this approach to organization, the CIF need not scrap labor and capital when it changes activities. This greatly frees resources within the firm to be utilized in different ways without losing the benefit of past investments.

The fact that CIFs are able to generate routinely improvements to products and production processes as a normal part of every worker's job suggest that a sharp division between those who use technology and those who innovate is the result of institutional impediments to worker innovation and the lack of a system for exploiting such a source of improvements.

The SM/MP firm's approach to control through restricting decision making authority and detailing the specific actions of workers precludes widespread innovation within the firm by lengthening the steps necessary to innovate. It may indeed be true that attempts for workers to innovate in an SM/MP organization would be too costly in terms of the needed effort to overcome these organizational impediments.

In contrast, the CIF organization supports and facilitates workers' contribution to product and process revision and improvement. Safeguards against misdirected changes are provided by guidelines detailing the methods to be used in initiating change. In this manner, CIF organizational characteristics can be viewed as a means for providing operational flexibility and malleability, while insuring high performance standards. These organizational characteristics insure diligent search, thorough consideration, and well-coordinated implementation of improvements.

The SM/MP firm's hierarchical structure is not intended to undergo continuous review and revision of its activities. Rather, it is intended to insure tight control of activities as they are, while benefiting from economies of scale and specialization of resources. The more rapid the pace of technological change, the greater the need for flexibility in response to changing circumstances, and the greater the need for customized company activities; the more the SM/MP firm will lag a CIF organizational approach. On the flip side, however, the CIF does not suffer a loss of economies of scale from its flexibility and malleability.[13]

The principle by which each type of firm is organized determines within each organization what is held constant and what is free to move. Within the SM/MP firm, specific actions are expected, but concern for the success of the company and others is not. An individual pursues their self-interest by increasing in specialized skills. Job security stems from the market value of specific skill sets to any firm. In the CIF, a worker's value is based on their ability to consider the overall needs of the company and to follow work guidelines, rather than on individual achievements. Each employee is encouraged to develop a cluster of both function specific and generalizable skills that can be applied to many jobs within the company.

IMPLICATIONS FOR ECONOMIC ANALYSIS

Economic modeling of technological competition - and the options and conclusions it provides - is based on premises of industrial organization with central decision making, pre-designated task content, and jobs based on specialized skills. In this schema; coordination, planning and change are controlled by a central authority. In this circumstance, the assumption that technological improvements are basically investment decisions and require specialists appears reasonable. It can also be assumed that technological decisions are relatively infrequent occurrences and need not be considered as a routine aspect of the production process. These assumptions, however, are inappropriate in an environment in which some firms have CI capabilities.

It is also suggested here that differences in industrial organization must be taken into account in assessing the costs and benefits of using CI methods. These organizational differences, because they greatly impact C/B assessments, affect the analysis of technological strategies and outcomes. Typically organizational factors are treated as outside the scope of economic analysis. However, these institutional differences diverge sufficiently from economic assumptions to merit explicit consideration when analyzing the options available to firms and the resulting competitive outcomes.

THE TRANSFER OF CI TECHNOLOGIES

In examining the transfer of CI technologies (and determining whether CI technologies can be transferred separately or as a bundle); the criterion, based on this analysis, would be whether CI derived technologies or the improvement creating technology itself are being transferred. CI derived technologies may or may not be of value to an organization. However, only the ability to generate and implement performance improvements continuously would keep pace with CIFs over the long run.

Whether local variants are "valid," from a competitive standpoint, ultimately depends on the competitive abilities of variants, rather than on the reasons for their existence. Thus, in evaluating adaptations, partial transfers and local variations of CI technologies; the criterion for success is how effectively the new organization can generate and implement improvements to its activities. As demonstrated in this analysis, change in a centrally controlled organization is awkward at best. The most to which such an organization can aspire is to perfect its use of externally developed technologies. Over time, however, its methods are likely to be outmoded by firms able to generate a continuous stream of improvements.

Considering the prerequisites necessary to improve continuously, however, it is conceivable that the adoption of CIF developed technologies could function as a stepping stone to developing CI capabilities. This is because the use of these technologies requires the mastery of statistical process control and other analytic methods that serve as the backbone of the CIF's technology development system. These techniques also provide the discipline and skill among workers and generate the detailed process information that is key to kaizen and concurrent design methods.

What these technologies omit, however, is the inventive process itself and what it requires of an organization.

CONCLUSIONS

Based on the examination of the organizational demands of technological change, it is concluded here that CI requires discipline; a system for generating, evaluating, implementing and monitoring innovation on a widespread basis within the firm; and a minimization of bureaucratic impediments to change. It appears that technological capability (both in technology adoption and innovation) is best served by organizations with diffused management and decision-making authority, superb coordination of activities, articulated methods of innovation, and high levels of discipline in worker behavior. In examining the differences between conventional industrial (SM/MP) organization and the CIF, we can make some general observations about the characteristics of these firms in a context of rapid change. The CIF has reduced organizational impediments to change by distributing the authority and skill to innovate to those directly involved in production activities. This and focusing workers on the smooth integration and coordination of activities has given the CIF admirable organizational responsiveness and malleability, and provided it with what Cole and Mogab (1994) have termed as economies of flexibility. In contrast, the SM/MP firm's focus on specialization to achieve economies of scale has lead to a greater need for centralized authority and planning and has made the firm's resources less malleable. This, in turn, has encumbered the SM/MP firm with a more hierarchical and functionally disjointed organization, making it both cumbersome and rigid in the face of change.

ENDNOTES

[1] This essay is based on work conducted while a graduate student at the University of Tennessee Knoxville and while a doctoral fellow at the United Nations University Institute for New Technologies (UNU/INTECH), Maastricht, The Netherlands. The author wishes to thank the faculty of INTECH, Jack Mogab, and William Cole for their many comments and suggestions on this and earlier incarnations of this work.

[2] This could occur either through investment by the firm in separate R&D initiatives or already established technology.

[3] Since much of the technology in use in western industrial powers is of indigenous origin, it may be that we are facing the first significant innovation to industrial organization to arise from a significantly different culture. In this case, much of the analysis of technological transfer problems from development economics literature may be appropriate here. As with any technology, CI technology affects the winners and losers in an economy, creating further resistance to change even to a recognized superior technology. This effect is compounded in the case of continuous improvement, as an organizationally based technology requiring change at the highest levels of an organization. This is explored later in this essay.

[4] See Fagerberg (1994) for a commentary on the role of stylized facts in the economic analysis of technology.

[5] The basis for this approach is provided in Schramm (1994).

[6] Imai (1986) provides a detailed explanation of the analytic tools used in developing kaizen innovations. Watanabe (1991) provides documentation of the significant scale and magnitude of such innovation activities in CIFs. Clark and Fujimoto (1991) describe the concurrent engineering approach used by Japanese auto manufacturing giants in product design.

[7] The characterization of the CIF used here is a composite of characteristics observed in a number of firms and is intended to represent an ideal. Individual firms vary in the degree to which they match this ideal.

[8] Should a company succeed in transferring techniques without transferring the technology creating components of CI organization, it is not necessarily out of the running for future CI capability. Much of the debate on CI transfer is akin to the question of whether a glass should be considered as half full or half empty, as mastery of the discipline and analytic capabilities required for CI are inherent to some of the techniques developed by CIFs. Mastery of these techniques,. therefore, may be able to function as a first step in building an organization capable of CI.

[9] While we have stated that CIFs are able to implement a continuous stream of both product and production improvements, we focus discussion on production improvements for brevity's sake. Product improvements can be treated as closely related to (and a derivative of) production process changes, as product characteristics are shaped by the characteristics of production inputs and processes.

[10] This might include product quality, timeliness of delivery, etc.

[11] While, by convention, economic analysis treats supply and demand as separate functions, production performance of firms greatly contributes to the broad spectrum of factors shaping demand for a product. In this manner, a product's supply and demand are interrelated.

[12] Suboptimal could be interpreted to mean that we adhere to the belief that there is an optimal production system available to the firm, if it can only discover it. This is not the sense in which we use the word. Rather, suboptimal performance refers to the existing production methods, but with the caveat that they are not functioning in a smoothly integrated fashion due to the recent disruption of change. This suboptimal performance is described by the author elsewhere as a performance gap and represents a negative aspect of learning curve behavior (Schramm 1994).

[13] Cole and Mogab (1994) explore this in depth.

REFERENCES

Arrow, Kenneth. "The Economic Implications of Learning By Doing." *Review of Economic Studies* 29 (June 1962): 155-173.

Best, Michael H. *The New Competition: Institutions of Industrial Restructuring.* Cambridge, Harvard University Press, 1990.

Clark, Kim B. and Takahiro Fujimoto. *Product Development Performance: Strategy, Organization, and Management in the World Auto Industry.* Boston, Harvard Business School, 1991.

Cole, William E. "Competitive Economies and the Economics of Competition." In *Competing Globally Through Customer Value* (Michael J. Stahl and Gregory M. Bounds, eds., pp. 14-31). New York: Quorum Books, 1991.

________ and John Mogab. *The Economics of Total Quality Management.* Cambridge, Mass.: Blackwell, 1995.

Fagerberg, Jan. "Technology and International Differences in Growth Rates." *Journal of Economic Literature* 31:3 (September 1994): 1147-1175.

Imai, Masaaki. *Kaizen: The Key to Japan's Competitive Success.* New York: McGraw- Hill, 1986.

Ishikawa, Kaoru. *Guide to Quality Control.* White Plains, New York: Kraus International Publications, 1987.

Monden, Yasuhiro. *Toyota Production System: Practical Approach to Production Management.* Tokyo: Industrial Engineering and Management Press, 1983.

Schramm, Stella. "Beyond Firm Learning: The Emergence of a Concurrent Model of Technological Change." Doctoral dissertation, The University of Tennessee, Knoxville, 1994.

Smith, Adam. *An Inquiry Into the Nature and Causes of the Wealth of Nations.* London: J. M. Dent & Sons, Ltd., 1910, Volume 1; originally published in 1776.

Watanabe, Chihiro; Irawan Santoso; Tjahya Widayanti. *The Inducing Power of Japanese Technological Innovation.* London: Pinter Publishers, 1991.

4

THE INDUSTRIAL DISTRICT AND CONTINUOUS IMPROVEMENT

Steven Wilson
United Nations Industrial Development Organization

A growing research effort over the past decade has contributed to a new view of the organization of the firm, innovation and competition (see Piore and Sabel 1984, Best 1990, Cole and Mogab 1995). This inquiry has been prompted in part by the rapid rise to economic predominance of Japan and the so-called "Asian Tigers" of the Pacific Rim and the seeming inability of conventional theory to provide a consistent explanation for this phenomenon.[1] Consequently, several researchers are now seeking in the organizational form of the firm the very sources of competitiveness and technological innovation.[2] A production-based analysis of competitiveness and technological change is now emerging which is a significant departure from the market-based methodology of mainstream economics.

This new view can trace its origins to the work of the pioneers in statistical quality control. W. A. Shewhart, his student W. Edwards Deming, as well as Joseph M. Juran and others developed tools which contributed to the foundations of a new analysis of production which would find no analog in the standard economic theory of the firm. While the earliest efforts in advancing this new system were largely North American in origin, the greatest application and continued refinement of this systemic approach has been in Japan. In very simplified terms, this methodology rests on the ceaseless and disciplined application of a range of improvement tools to increase the quality of products and the robustness of productions systems. For this reason the term "continuous improvement" is often used to describe this approach.

Central elements of the research into the process of continuous improvement (see Cole and Mogab 1994 and Schramm 1994) may offer analytical insights that can enrich the inquiry into the "industrial district" literature (see Humphrey 1995, 149-

162), and Sengenberger and Pyke, (1991). The industrial district will be discussed in some detail in section III. This work is a departure from most analyses of industry in developing countries because of its examination of practical institutional arrangements for small to medium sized enterprises (SMEs).[3] A growing empirical literature suggests that key aspects of the industrial district are viable in developing countries.[4] Significantly, this literature highlights the concrete role of manufacturing and industry in the process of economic development, a focus that has largely faded in much recent policy and theoretical discussion.

THE CHANGING FORTUNES OF INDUSTRIAL DEVELOPMENT[5]

The first generation of post-war development economists was unambiguous in their support for industrial development. It was argued that industrialization could generate a wide range of positive externalities, sustain high growth, and absorb surplus labor.[6] Industrial linkages would provide the rationale for investments that might not otherwise be made and would stimulate new and dynamic growth paths. Accepted theory lent support to state coordination of large scale public investment in order to lay the groundwork for the "big push" of modernizing industrial development.[7] The industrial success of the Soviet Union - equal to the task of defeating Germany in the Second World War -further convinced many observers of the value of investment planning and state-led industrialization.

While trade was recognized as an important "engine of growth," an influential view held that free trade would consign developing countries to the position of raw material suppliers. This belief combined with the theory of the big push to lay the foundations for infant industry protection in developing countries.[8] The outcome of this was the widespread adoption of various programs that came to be known as import substitution industrialization (ISI). But by the 1970s the concept of industrial development had lost its luster in both theory and policy. Analysis shifted away from the issue of industrial externalities and linkages (Stewart and Ghani 1991). And benefit-cost studies emphasized the effect of tradable production upon international trade rather than domestic resource use.

This effectively ruled out the impact of domestic industrial linkages and externalities (e. g,. Little and Mirlees 1969, 217). Increasingly, industrialization in developing countries was seen as induced by a range of trade and domestic distortions and wet-nursed by ongoing state support (see Little et al. 1970). By the 1980s the issue of positive industrial externalities - themselves prima facie evidence of market failure - did not elicit the attention of the most influential economists.

However, it was the widespread failure of ISI regimes that provided the strongest vindication for the opponents of industrial development policies in developing countries. Infant industry protection allowed domestic elites - including many in government - to amass fortunes without creating a competitive industrial base. Domestic joint ventures with transnational firms often operated as assembly operations within protected markets. Few of these adopted new technologies or production techniques (Best 1993). Paradoxically, most ISI regimes actually increased the vulnerability to fluctuations in foreign exchange receipts (Colman and Nixson 1985).

After years of stagnation and rent-seeking behavior associated with ISI, support broadened for market-oriented policies among those across the political spectrum. By the 1980s politicians across the developing world were elected upon platforms advocating sweeping market liberalization. Market reforms were embraced as the treatment for ailing industries in the developing world and state support was cast as the principle malady. In a profound departure, several developing and newly industrializing countries unilaterally reduced trade barriers well in advance of the cuts mandated by the Uruguay Round of GATT.

A core tenet of development thinking - that poor economies differed from the developed in important ways - receded in the face of the universalism of the market or neoliberal approach. Differences between the "developed" and the "developing" countries were now largely seen as the consequence of inappropriate macroeconomic and trade policies and poor governance rather than of fundamentally dissimilar economic systems. Sub-optimal economic performance could now be traced for the most part to government interference in the operations of the free market. It followed that this market was seen to obey the same laws everywhere.

Once market interference was removed, developing countries could now expect patterns of specialization in both trade and manufacturing to emerge based on their relative comparative advantage. Trade was reaffirmed as an "engine of growth," but not the asymmetric trade favoring the "South" over the "North" as enshrined in GATT of 1964. Rather, the optimal path to the economic "convergence" of nations would be based upon free trade and a sweeping liberalization of a nation's economy (see Sachs and Warner 1995). At the level of the firm the diffusion of "best practices" and organizational structures would obey the same logic as any other technological choices made by entrepreneurs in a competitive market place. By the 1990s the neoliberal view had emerged as the predominant approach to development and growth.

The sudden and unexpected dissolution of the Soviet bloc produced many eager acolytes of market economics, many of them former Marxists fleeing industrial sclerosis and failed planning. Structural adjustment packages applied in developing countries now downgraded the role of industrial externalities and focused on price liberalization, privatization, and fiscal and monetary austerity. At the same time the growth in the global share of manufacturing value added and manufactured exports from the developing world continue to climb steadily upward. These successes in trade and industrial development provide ample evidence to many of the superiority of unfettered markets.

However, the lack of a tractable and consistent explanation for major trends of the late twentieth century troubles many economists. For example, the superior growth rates of Japan and the Asian Tigers, and their capacity to challenge and dominate in a number of sophisticated sectors, can no longer be ignored nor dismissed as an aberration. Their rise was not predicted nor adequately explained by traditional theory. And their success is based in part on the development of those very areas for which they had no early comparative advantage. Despite the insistence by some that these nations are "special cases" which cannot be generalized, their continued success and growing weight in the global economy poses a challenge to the universality of the neoliberal world view.[9]

Moreover, in spite of the stunning successes of a small number of newly industrializing countries, a much larger number of developing countries continue to stagnate. The dismantling of ISI regimes, widespread market liberalization, the imposition of structural adjustment programs and the exposure to international competition have brought mixed results. Branch plants of transnationals in developing countries have often shut down or relocated in the wake of exposure to international competition instead of metamorphosing into viable and competitive operations. And national businesses often find themselves ill-prepared to compete against North American and European firms as well as the group of Pacific Rim corporations which have highly developed production and export capabilities.

And low wages continue to provide the basis for the acceleration of globalized production based upon the practice of "slicing up the value chain." This will likely spawn greater opportunities for free trade zones and export platforms based on cheap wages across the developing world.[10] This practice is driving the impressive increases in manufactured exports and manufacturing value added in developing countries. Indeed, the fact that manufactured exports from the developing world are causing concern in the most advanced market economies is proof of the stunning reversal of old trading relations. Such a development is cited by many as proof of the gains from trade based upon comparative advantage.

However, much of the investment in developing countries manufacturing is located in plants where labor-intensive operations are broken down into simple and repetitive operations. This specialization is made possible by designing products and production processes so that constituent parts of the product can be assembled or manufactured based on the optimal mix of location, labor cost, labor quality, and other enticements offered by host governments. Importantly, this design function is most often performed in a place different from the actual production site. Direct labor costs are minimized across the production chain and the employer can move standardized operations to other locales with relative ease. "Footloose" producers can nullify any upward wage pressures by relocating elsewhere. This has led several host governments (e.g., such as the case of Mexican policy vis-à-vis the maquiladora industry) to pursue low wages in order to attract and keep labor-intensive operations (Wilson 1992).

Importantly, in labor-intensive plants in which workers perform highly specialized and simplified operations, the opportunities for on-site continuous improvement are often diminished. Workers do not gain much experience in such an environment and management sees little incentive to invest in worker training and development. High turnover reinforces the drive to break the production process of a product into its smallest, irreducible parts and simple and repetitive procedures mean a new employee can be substituted with little or no loss to productivity.[11] Low wages continue to attract investment, but efficient forms of organizational improvement which stress the removal of a wide range of non-value adding activities which generates far greater increases in high quality output than does adding labor however low the wage.[12]

Outsourcing labor-intensive components of the value chain will generate more employment in low wage assembly operations, especially in choice locations located within or near growth markets. This is based upon comparative advantage.

But where ISI programs failed to upgrade low wage operations with heavy protection and strong state support, the free market approach offers no mechanism for converting "maquiladora"-type enterprises into locally viable bases of production. Where import substitution schemes relied on market controls and regulation to spur production, the free market approach sees perfect markets as promoting production. Both approaches focus on the role of markets and overlook the importance of production. However, neither view has provided the conceptual link between markets and production (Best 1993).

Meanwhile a new wave of manufacturing exporters led by China (a virtually autarkic Communist economy a generation ago) has fueled social concerns in the developed countries and stirred deep debate (see Wood 1994). Wide ranging turbulence in the global economy and the anomalous growth of the Pacific Rim economies suggest that any economy is in or could fall into a low level equilibrium trap (Sabel 1993 6-8).[13] New theories of growth and trade, solidly within the framework of standard theory, have also called into question assumptions underlying laissez-faire and the existence of equilibrating market forces which allow the weak to catch up with the strong nations.[14] And other work reopens interest in a wide range of externalities, including industrial externalities.[15]

INDUSTRIAL DISTRICTS AND CLUSTERS

The industrial district literature is one attempt to bring the role of production and industrial activity in developing countries back into the broader study of economic and industrial development. This represents a return to a long tradition in which industrial activities were seen as central to the process of development. The term "industrial district" was first used by Alfred Marshall a century ago to describe groupings of similar firms which gave rise to what he called "external economies" (Marshall 1919). The industrial district literature is based primarily upon the European experience, but increasing attention is being turned to the applicability of this approach for a large number of SMEs in developing countries. This is an important issue given that most SMEs are currently not competitive and the global wave of liberalization threatens many with extinction.

While the subject of small-scale industry has received a great deal of attention in recent years, analytically it remains one of the least explored areas in development studies (Schmitz 1990, 257). Work on industrial districts and clusters differs from most conventional studies of small-scale industry which emphasize the formidable constraints arrayed against small enterprises. The socially desirable role that SMEs play in employment generation is often highlighted, but this view has always collided with the main rationale for larger plant size: the exploitation of greater scale economies. The verdict is often that "small may be beautiful," but big is more efficient. Capital-starved and technologically primitive firms therefore require radical upgrading, increased scale, consolidation or outright euthanasia.

As a point of departure from conventional studies of SMEs, the industrial district literature examines the small firm as part of a synergistic group of enterprises. Where an individual firm specializes in one part of a production chain, a small firm is said to be able to take advantage of a variety of external economies that arise from linkages with others within the chain. Intra-firm improvements are spread to others within the district and are in turn reinforced by the institutional and cooperative mechanisms within the industrial district. Importantly, such specialization is thought to give SMEs a high degree of flexibility which can be used to respond to niche markets - or to supply other firms. In this sense "smallness," contrary to the conventional view, can be the source of competitive advantage. Given the poor performance of both small and large-scale enterprises in many developing countries, this newer literature merits closer study.

The idea that small firms in traditional sectors can thrive under "bounded competition," competing in some areas and cooperating in others - has gained currency in the literature. This work draws on a diverse strand of research and can be decomposed into three broad but interrelated areas: the macro level of changing forms of production and competition, the micro level which is concerned with improvements inside the firm, and at the inter-firm level in which a variety of linkages and networks are examined.

At the macroeconomic level investigation centers upon the shift from the dominant mass production system in which stable markets, factor-cost reductions, and scale economies were the key variables to more rapidly changing and segmented markets, products, and processes where flexibility and innovation are the key competitive metrics (Piore and Sabel 1984 and Best 1990). In this context the literature examines industrial policies designed to facilitate the process of change and adaptation with a growing focus on the developing countries. The exigencies posed by new forms of competition shape the appropriate policy responses and determine in large part the strategies of individual firms.

At the micro level, authors have drawn upon a large literature that analyzes organizational change and improvement methods inside the individual firm. Piore and Sabel examined smaller firms in Italy and elsewhere in Europe as well as small Japanese firms in extensive supply chains. These authors were particularly impressed by small Italian producers who focused on one aspect of a production chain, using specialized machinery and labor, and who were geographically close to other producers and suppliers in the same chain. Other analysis draws heavily on the study of reorganization and improvement within large firms particularly in the Japanese case. A growing number of analysts are examining the efficacy of such methods in smaller firms in the developing countries.

However, it is the analysis of inter-firm behavior among small firms that occupies the lion's share of the industrial district literature. Here the role of industrial districts and dense webs of linkages and networks are examined. Inquiries go beyond single firms to study the dynamics of clusters of several interconnected firms. The original model for this approach is that of the "third Italy" model of the Emilia-Romagna region. In the late 1970s the traditionally prosperous northwest (first Italy) and the impoverished south (or second Italy) were both mired in slow growth and high unemployment. In contrast, the Northeast and the center of the country, or the third Italy, registered impressive industrial growth.

This success was based on a large number of smaller manufacturing firms arranged in what have been called "industrial districts" (Becattini 1990). Alfred Marshall first used the term to describe configurations where "external economies" arose due to an increasing division of labor among firms that were located close to one another and concentrated in the same sector. The third Italy model has the search for similar industrial districts in a number of developed country regions and provides a schema for exploring for similar configurations in the developing world (Schmitz 1995). While there are many variations among industrial districts, the following four main elements are thought to be fundamental characteristics:

- small and medium sized enterprises which are clustered in proximity to one another and are sectorally specialized,
- a set of forward and backward linkages among economic agents, based on market and non-market exchanges of goods, information and people,
- common cultural and social background linking the economic agents and creating a behavioral code - a code which is sometimes explicit but often implicit,
- and the existence of public and private local institutions acting to support the cluster.

A fundamental premise is that small firms arranged in clusters can be as viable, or in some cases more so, than larger firms. By clustering it is argued that firms can gain economies of scale and scope as well as increased production flexibility through specialization and inter-firm cooperation. The Italian experience suggests that small firms can be highly competitive internationally in traditional products such as shoes and textiles without reliance on low wages and poor labor standards (Humphrey 1995). The emergence of prosperous industrial districts in high-wage locations in other locations in Europe increase the interest that such an arrangement might increase competitiveness elsewhere (Sengenberger and Pyke 1991; Schmitz and Musyck 1994).

While making the same or similar items in vicinity to other producers may constitute a cluster, this in itself brings few benefits. A grouping of fashionable boutiques on a city street or the concentration of sweatshops making similar clothing do not constitute a cluster in a district. They share nothing but the same line of business. It is the balance between competition and cooperation that is thought to underlay the viability of an industrial district. Cooperation helps to overcome the disadvantages of smallness while competition is seen as promoting the dynamism and flexibility that larger firms lack.

A number of developments must occur for a cluster of small firms to reap the competitive gains which have been loosely termed "collective efficiency" (Schmitz 1990). Among these conditions are:

- a division of labor and specialization among small producers,
- the provision of specialized products or services at short notice,
- the emergence of raw materials, components, or new and used equipment suppliers,
- the emergence of agents to sell to national/international markets,
- the emergence of specialized services in technical, accounting, financial matters,
- the emergence of a pool of wage labor with sector-specific skills,

- and the emergence of consortia for specific tasks (e.g., exporting) and of associations providing services and lobbying for its members

Collective efficiency is broader than Marshall's "external economies," as it comprehends joint action through producers' associations or local government to tackle common problems such as site selection, services, training, and so forth. Collective efficiency does not preclude competition, which can be intense among members of a cluster. Vertical relationships in which firms buy products and services in the market or through subcontractors can be governed through strict price bargaining, or can come via some form of strategic collaboration. And horizontal relations among sectorally clustered firms are often characterized by outright competition.

But a main point to be culled from Marshall's external economies and Schmitz' concept of collective efficiency is that clusters in industrial districts can give rise to benefits that are not strictly "of the market," i. e., determined by price. Schmitz suggests that while there may not be a compelling reason for one firm in isolation to cooperate, the gains accruing to clustered firms can urge self-interested agents to engage in joint action on a variety of activities even with competitors. This would include shared learning with the consequence of diffusing innovation and overall increases in productivity.

Indeed, one of the more provocative claims made in the literature is that intra-firm reorganization and improvement is most effectively supported when firms are seen as parts of networks rather than isolated units. Some empirical studies conclude that firms need to reorganize internally and redefine their links with other firms simultaneously, each being a precondition for the other. Messner's (1993) study (cited in Humphrey 1995) of the Chilean wood industry identifies a sequential process of improvement. First, firms incorporate new methods and training internally, and then they establish strategic inter-firm groups. These groups establish horizontal and vertical structures which improve their overall capacity and create internal pressures for improvement through the imitation and propagation of innovations.

The literature of what we have loosely termed the industrial district is an attempt to establish a viable role for the small firm in the context of new forms of global competition. Smallness need no longer be viewed as a mere precursor to largeness or to extinction. But instead, it can be seen as a possible source of strength in the appropriate institutional milieu. Moreover, the industrial district literature offers a possibility for the sustained growth of smaller firms without invoking the market clearing assumptions of conventional economic theory.

However, the literature on the industrial district, and the constituent concept of clusters, do not approach the degree of generalization required of a "model."[16] Firm types, their arrangement vis-à-vis other firms in a cluster, and the kind of institutional involvement can vary widely across various agglomerations identified as "industrial districts." The experience of the "third Italy" is quite different from that found in other places such as Baden-Württemberg, the Salzburg region, "Route 128" near Boston or the Silicon Valley. In certain areas identified as industrial districts, family-owned firms are clustered with others of similar size and capitalization. In other instances, specialty craft shops operate in proximity to one another and supply large regional or even multinational producers. The importance of "shared socio-cultural identity that facilitates trust" is often cited, yet this does not serve as a guide for the promotion of the industrial district in developing countries or elsewhere. The industrial district literature remains more a set of stylized facts than a fully developed model or theory of industrial organization. Yet a growing body of empirical work suggests that the concept can be applied in the developed country setting. For many analysts the approach is a viable departure from the orthodox view which stresses the virtues of atomistic competition among small and isolated firms at each and every stage in the production process. Certain forms of cooperation and facilitating institutional arrangements are seen as more effective ways of transmitting improvement across small firms. But what kinds of improvement process are implied by the industrial district model?

INDUSTRIAL DISTRICTS AND CONTINUOUS IMPROVEMENT: ONE AND THE SAME?

The literature on the industrial district makes numerous claims regarding the improvement-enhancing synergy of clustered and interlinked firms. The spread of "TQM" and "Japanese quality and management methods" are thought to be made easier in firms clustered in an industrial district. Indeed, some authors seem to imply that the institutional arrangement of the industrial district and "improvement methods" are somehow synonymous. This confusion is traceable to the fact that advantages identified with the industrial district, on the one hand, and TQM or continuous improvement on the other, stem from different sources.

Piore and Sabel (1984) developed the concept of "flexible specialization" which is a major conceptual underpinning of much of the industrial district literature. Their work examined a diversity of firms, from SMEs arranged in industrial districts, to so-called "solar firms" epitomized by Boeing with its web of suppliers, to federated firms such as pre-war "zaibatsu" in Japan, and to a range of "workshop" factories (Piore and Sabel, 265-8). The large-scale variant of flexible specialization involves firms that have decentralized and taken advantage of labor and product specialization or those which use specialized suppliers. The small firm variant finds its best-known expression in the industrial districts of third Italy. Importantly, the process of improvement is seen to be analytically indistinct in both the large and small cases.

The essence of Piore and Sabel's analysis is that the passing era of mass production and the fragmentation of markets present "flexibly specialized" small firms with opportunities to exploit new forms of flexible machinery and work organization. By specializing in certain parts of the product chain and being arranged in an industrial district - with its supportive institutions and cooperative elements, small firms can rapidly and flexibly respond to evolving market niches. In the Italian case, the coordination of production, deliveries, learning and improvement is done through a variety of informal and non-hierarchical arrangements for reaching consensual decisions.

For Piore and Sabel and others in this tradition, one advantage of flexible specialization within an industrial district is the return to a craft tradition based upon highly specialized labor. Highly specialized workers using flexible machinery to respond rapidly to the market are one of the advertised strengths of this system. Indeed, innovation within the system depends on the degree of specialization of labor.

However, the issue of specialization and intra-firm improvement is given far less attention that the role of relations among small firms (Humphrey 1995). Yet there is little in the literature which examines how improvements or new methods area actually transmitted or any insight into differences among firms to generate and absorb new methods and improvements. Terms such as "improvement" and "innovation" are used interchangeably with no way of knowing if progress is incremental, continuous, or discrete. Any differences in the way firms handle production-specific information are also downplayed. Where the roles of managers are discussed, it is typically in the context of their relations to other firms within a district.

Recent studies of continuous improvement or TQM within the firm, such as that of Cole and Mogab (1995) and Schramm (1994), are based on a different type of improvement process. This work examines the key characteristics of the archetypal Japanese "continuous improvement firm" (CIF). Contrary to the industrial district, this work reveals that the competitive strengths of the CIF are not based on deep specialization and division of labor. In the CIF tasks may be specialized, but workers are generalized. And it is the cross-functional perspective of the generalized labor force which helps promote potent improvements.

This has important implications for the transfer elsewhere of tools and practices developed within the CIF. These practices were developed by users interested in improving every aspect of the production process from research and design to customer service after the purchase. Competitive abilities such as JIT manufacturing are the outcome of a tightly linked process that uses production-specific information to generate ongoing and incremental improvements. Being developed within a comprehensive systems perspective means that such tools are often difficult, if not impossible, to transfer to firms that are not managed from such a view.

Therefore, it is unclear how a system of small firms based upon highly specialized labor can effectively exploit certain tools (e.g.. JIT) developed for use within a systems-based, cross-functional approach to improvement. In addition, we have little idea how information and learning is transmitted across such a system of small firms in a district nor the character or extent of feedback loops into future rounds of production.

The concurrent approach to technological change within the CIF has allowed this firm to fully exploit firm-specific information and bring about an important form of endogenous change. Forms of "on line learning" have given the CIF an ability to create and maintain a performance gap over rivals. This requires the precise reduction of variation in the production process, strong control and discipline, and a well developed level of communication (Schramm 1994). How small firms organized in an industrial district could effectively implement such a concurrent approach is an important area for research.

Within the CIF emphasis is placed on the role of top managers operating across functions and the benefits which accrue from this practice in terms of systemic improvement. In the industrial district the presumed role of the manager is to use specialized craft-type labor to produce flexibly what is required in the production chain. The mechanisms by which joint decisions are made among a large number of managers in separate plants and the use of production information by managers, and to feedback into a sustained process of improvement are undeveloped topics in the literature.

As stated, this literature on the industrial district sees no important differences in the improvement process of a production system run by one top manager in a large enterprise and those of several small and specialized firms, each headed by their own manager. Such an assumption assigns little importance to differences in management and suggests that the diffusion of improvements should be more or less uniform across different sizes of firms. Even where there exists a common cultural and social background, such an assumption may not be realistic. But as research into the continuous improvement process has shown, such differences in the diffusion of improvements do exist and can persist (see Cole and Mogab 1995 and Schramm 1994).

The industrial district is based on the assumption that specialized labor, new types of flexible machinery, and firm networks will automatically promote improvement. Moreover, shared cultural and social norms are seen as essential in promoting forms of cooperation. This literature and concept will greatly benefit from an analysis of the nature of the improvement process itself. While supporters of the industrial district are quick to point out that this model is a break with "fordism" and "mass production" - they may be unwittingly falling back on two of the core rationales of the "old paradigm": specialization and the division of labor.

And the issue of shared socio-cultural norms do not help us very much in promoting districts and clusters elsewhere.[17] Unusual cultural and historical characteristics, perhaps in the third Italy, might make this unique approach quite potent. It may not be so elsewhere. Further research must focus on what elements can be distilled from the Italian experience, among others, and be implemented elsewhere. It is possible that certain aspects of continuous improvement are be culturally "embedded" in the Japanese experience, yet a wealth of empirical work suggests that a large number of these tools are applicable anywhere. Similar work should be done in support of an industrial district model. The outcome could be a more robust model and a greater vehicle for policy aimed at increasing the competitiveness of SMEs in developing countries.

ENDNOTES

1 Conventional explanations for East Asian success revolve around rapid physical and human capital accumulation. See Collin and Bosworth (1996), Young (1995), and Kim and Lam (1994).

2 Competitiveness - or international competitiveness - is a widely used term for which no concise or common definition exists. Loosely speaking, the international competitiveness of a firm - or industry - can be assessed by its relative ability to sell its products on international markets. In the standard literature the term competition refers to specific conditions under which production and trade take place. These conditions are determined by the particular market structure present. Therefore, a particular kind of competition is characteristic of a specific kind of market structure. And it is the form of competition that determines in large part the international competitiveness of a country in a given industry (see Forstner 1995, 181-204). However, this standard view omits any consideration of the role of different forms of production at the level of individual firms in determining competitiveness. Instead, it is clear that the emphasis is on market structure. The investigation into the "black box" of production marks the departure from traditional theory of the theorists of the "new competition."

3 There are a large number of definitions for "SME" used by various organizations. None has gained majority acceptance. Some definitions stress the number of employees, while others emphasize measures of profit or capital utilization. The range of these is considerable with some definitions comprehending shops with a handful of employees, and others consigning anything below twenty or so employees to the category "micro-enterprises." UNIDO defines an SME to be a privately owned domestic enterprise with between five and 200 employees.

4 See for example the January 1995 issue of *World Development* which was devoted to this subject.

5 The common intermingling of the terms "industrialization" and "industrial development" can cause difficulties. Industrialization refers to the growth of manufacturing and processing activities within an economy and can be viewed by metrics such as manufacturing value added. Industrial development, on the other hand, is meant to connote the role of industrialization as part of a broader process of economic and social development. The latter concept is explicitly concerned with the role of economic linkages and externalities in the industrialization process.

6 See Rosenstein-Rodan (1943), Singer (1950), Lewis (1954), Mydral (1957), and Hirschman (1958). The seminal work in the empirical literature on industry externalities remains that of Kaldor (1967) who devised and tested the hypothesis that the manufacturing sector output growth generates externalities which raise this sector's productivity growth. Other approaches have used of input-output models to estimate inter-sectoral linkages (Yotopoulos and Nugent 1976, 259-285).

7 In Hirschman's elegiac article on development economics (1981) he suggested that one central tenet of development theory had been the belief that economies trapped in a low-level equilibrium were fundamentally different from those that

were largely self equilibrating. This acknowledged difference, along with Keynes' analysis of capitalism and other work such as Gerschenkron's analysis of the role the state and Lewis' work on dual economies, all underlined the need for state support (Keynes 1936, 1980; Lewis 1954; Gerschenkron 1962).

8 Sachs and Warner (1995, 13-14) suggest that the greatest impetus to state-led industrialization policies after the Second World War was the fact that the world trading system itself was in shambles. Given that almost none of the richer countries had convertible currencies or low external tariff rates, much of the "export pessimism" of developing countries was justified.

9 The nascent economic powers of the Pacific Rim have been identified by many as further proof of the benefits of competition and trade. Certainly the economies of the region have excelled in trade and are characterized by a high degree of domestic competition. However, this competition differs in significant respects from that described in the standard economic model (Cole and Mogab 1995). Other observers suggest high growth countries are largely explained by rapid and massive capital accumulation along with upgraded human capital (see Mankiw 1995, 275-326). While such developments are characteristic of several East Asian economic powers, the manner in which such accumulation and skill upgrading occurred seems to make a better case for state-led industrialization than laissez-faire. Certainly the theory of comparative advantage would not have been much of a guide in predicting that Japan, virtually devoid of raw materials, would become the world's most efficient and profitable steel producer. Several attempts to explain the Asian "market miracles" within the confines of standard theory have metamorphosed into overlapping waves over the years. The lack of a consistent explanation of this phenomenon casts doubt on any long-term logic of world-market equilibrium.

For others the experience of Japan and the Asian Tigers remains a special case. In this view, systems of keiretsu, chaebol and the like are seen as largely unique to specific cultures and historical experiences. Variants of this view suggest that these countries developed while pursuing neo-mercantilist policies under the aegis of a strong state that was intertwined with domestic industrial and financial interests. Export success depended upon relatively open markets elsewhere, particularly in North America. The Cold War and the U.S.-led policy of Communist "containment" placed these trading states in a favored and protected geopolitical position. In short, these countries did not have to develop under the competitive rules which face contemporary developing countries and therefore their experiences are not applicable, nor particularly relevant, to other nations.

Yet this argument fails to explain the continued growth and competitiveness of the East Asian powers. And in any case, interference with the free functioning of the market in East Asia should, according to received theory, have impeded the very progress that these nations have enjoyed. Nor did theory help predict the emergence of economic dynamos based on knowledge-intensive industries. In short, the continued development and growth of these Asian powers can not be dismissed as a special exception to the rules. In such circumstances it is an open question as to which is in fact is the "special case," the East Asian experience, or the broad explanatory power of standard theory.

Moreover, if the experience of these countries is unique and not applicable to developing countries, it does not necessarily follow that the precepts of neo-liberalism are more universal. Yet this is implicit in structural adjustment programs applied across the globe. These policies have mostly failed to bring developing countries, particularly the least developed ones, out of low-level equilibrium traps.

10 Some observers have suggested that labor costs have become such a low share of total costs that low wages no longer represent an advantage in attracting investment. However, this is not the same as saying that labor costs have declined as a share of value added - they have not. Instead, what most people appear to mean is that because of the growing vertical disintegration of industry, the value added by a given manufacturing facility is likely to be a small fraction of the value of its shipments, and thus labor's share of value added is also a small fraction of total costs, the majority of which are dominated by the costs of intermediate inputs. However, with growing vertical disintegration of the production process, or slicing up of the value chain, there is likely to a greater incentive to relocate production to low wage platforms (see Krugman 1995, 327-377).

11 This approach can in some cases actually increase employee turnover by making the job incredibly boring and repetitious (Wilson 1992).

12 Such developments blur conventional measures of productivity. In the case of Japan, conventional thinking led earlier observers to predict a loss of Japanese industrial competitiveness because of rising wages. This analysis completely missed central issues of continuous improvement that did not hinge on the wage rate.

13 Turbulence in financial and currency markets also suggest a lack of agreement as to the causes of disturbances. One need only view the oscillations of the world's premier reserve currency - the U.S. dollar - and the weekly de novo theorizing as to the causes of its instability to question accepted tenets of standard theory. Likewise, confident predictions about the impacts of an appreciating yen on the U.S. trade deficit with Japan call into question market analysis that completely excludes new forms of production and competition.

14 For example, in the well-known neoclassical view, labor and physical capital as factors of production are subject to diminishing returns. The larger the given stock of capital, therefore, the smaller the corresponding output from an increment of capital. It follows therefore that investments should have a higher return in developing countries with smaller capital stocks than in the developed economies. Given the mobility of factors developing economies should attract investment and in time their growth rates should converge with those of advanced countries. Apart from well-documented Asian examples and a handful of other newly industrializing countries, many other poorer economies have become relatively poorer. Romer among others attempts to introduce knowledge as factor which is subject to increasing returns, and hence can stave off diminishing returns to factors such as capital (see Romer 1990, Lucas 1988, Mankiw 1995).

15 Proponents of the new growth theory, (Helpman 1992, Lucas 1988, Romer 1986, 1989, 1990) as well as the new theory of international trade (Krugman 1990, 1991; Rivera-Batiz and Romer 1991) have examined the role of several

types of externalities, including knowledge, human capital, and technology. Other literature has opened a reexamination of externalities in the industrial process (see Khan and Bilginsoy 1994 and Stewart and Ghani 1991.

[16] The ambiguity surrounding the many terms used by those who publish in this and related literature further cloud the issue. Flexible specialization, flexible production, post Fordism, post modernism, new industrial districts, networks, new industrial spaces, industrial milieu and other terms are often used to refer to very similar or identical issues.

[17] Current UNIDO projects promoting clusters where a tradition of collaboration of cooperation is lacking rely on more formal structures such as legally binding contracts signed by participants which describe their obligations to the cluster.

REFERENCES

Becattini, G. "The Marshallian Industrial District as a Socio-economic Notion," In F. Pyke, G. Becattini and W. Sengenberger, eds.,. *Industrial Districts and Inter-firm Cooperation in Italy.* Geneva: International Institute for Labour Studies, 1990.

Best, Michael. "The New Regulation." *Mimeo.* Vienna: UNIDO, 1993.

Brusco, Sebastiano. "The Idea of the Industrial District: Its Genesis." In F. Pyke, G. Becattini and W. Sengenberger, eds., *Industrial Districts and Inter-Firm Cooperation in Italy.* Geneva: International Institute for Labour Studies, 1990.

Cole, William E. and John W. Mogab. *The Economics of Total Quality Management: Clashing Paradigms in the Global Market.* Cambridge, MA: Blackwell Publishers, 1995.

Collins, Susan M. and Barry Bosworth "Economic Growth in East Asia: Accumulation versus assimilation." *Brooking Papers on Economic Activity* 2 (1996): 135-191.

Colman, David, and Frederick Nixson. *Economics of Change in Less Developed Countries.* 2nd ed. Oxford: Philip Allan Publishers, 1986.

Forstner, Helmut. "The International Setting." *In Policies for Competition and Competitiveness: The Case of Industry in Turkey.* Vienna: UNIDO, 1995.

Gerschenkron, Alexander. *Economic Backwardness in Historical Perspective.* Cambridge: Harvard University Press, 1962.

Hasegawa, Keitaro. *Japanese Style Management : An Insider's Analysis.* Tokyo: Kodansha International, 1986.

Hirschman, Albert O. *The Strategy of Economic Development.* New Haven: Yale University Press, 1958.

________. "The Rise and Decline of Development Economics." *In Essays in Trespassing.* Albert O. Hirshman, ed. Cambridge: Cambridge University Press, 1981.

Humphrey, John. "Industrial Reorganization in Developing Countries: From Models to Trajectories." *World Development* 23:1 (January 1995): 149-162.

Imai, Masaaki 1990 "The Kaizen Wave Circles the Globe." *Tokyo Business Today* (May): 44-8.

Kaldor, N. *Strategic Factors in Economic Development*. Ithaca: Cornell University Press, 1967.

Keynes, John Maynard. *The General Theory of Employment, Interest, and Money*. New York: Harcourt, Brace, and Javanovich, [1936], 1980.

Khan, Shahrukh R. and Cihan Bilginsoy. "Industry Externalities Revisited." *Kyklos* 47:2 (1994): 67-80.

Kim, Jong-I and Lawrence J. Lam. "The Sources of Economic Growth of the East Asian Newly Industrializing Countries." *Journal of the Japanese and International Economics* 8:3 (1994): 235-272.

Krugman, Paul. "Growing World Trade: Causes and Consequences." *Brookings Papers on Economic Activity* 1 (1995): 327-377.

Lewis, W. A. *Economic Development with Unlimited Supplies of Labor*. The Manchester School 22 (May 1954): 139-191.

Little, I.M.D. and J. A. Mirrlees. *Manual of Industrial Project Analysis in Developing Countries*. Paris: OECD, 1969.

Little, I.M.D., T. Scitovsky and M. Scott. *Industry and Trade in Some Developing Countries: A Comparative Study*. London: Oxford University Press, 1970.

Lucas, Robert E., Jr. "On the Mechanics of Economic Development." *Journal of Monetary Economics* 22:1 (July 1988): 3-42.

Mankiw, N. Gregory. "The Growth of Nations." *Brookings Papers on Economic Activity* 1 (1995): 275-326.

Marshall, Alfred. *Industry and Trade*, 3rd edition. London: Macmillan, 1972.

Messner, Dirk. "Búsqueda de Competitividad en la Industria Maderera Chilena." *Revista de la CEPAL* 49 (April 1993): 115-136.

Myrdal, G. *Economic Theory and Underdeveloped Regions*. London: Ducksworth, 1957.

Piore, Michael and Charles F. Sabel. *The Second Industrial Divide*. New York: Basic Books, 1984.

Romer, Paul M. "Endogenous Technological Change." *Journal of Political Economy* 94:5 part 2 (1990): S71-S102.

Rosenstein-Rodan, P.N. "Problems of Industrialization in Eastern and South Eastern Europe." *Economic Journal*. 53 (June/September 1943): 202-211.

Sabel, Charles F. "Learning by Monitoring: The Institutions of Economic Development." *Mimeo*. Vienna: UNIDO, 1993.

Sachs, Jeffrey D. and Andrew Warner. "Economic Reform and the Process of Global Integration." *Brookings Papers on Economic Activity* 1(1995): 1-118.

Schmitz, Hubert. "Small firms and flexible specialization in developing countries." *Labour and Society* 15:3 (1990): 257-275.

________. "Small Shoemakers and Fordist Giants: Tale of a Supercluster." *World Development* 23: 1 (1995): 9-28.

________and B. Musyck. "Industrial districts in Europe - policy lessons for developing countries." *World Development* 22:6 (1994): 889-910.

Schramm, Stella Sherrouse. *Beyond Firm Learning: The Emergence of a Concurrent Model of Technological Change*. Unpublished doctoral dissertation, The University of Tennessee, Knoxville, 1994.

Segenberger, Werner, and Frank Pyke. "Small firms, Industrial Districts, and Local Economic Regeneration." *Labour and Society* 16:1 (1991): 1-25.

Singer, H.W. "The Distribution of Gains between Investing and Borrowing Countries." *American Economic Review* 40 (May 1950): 473-85.

Stewart, F. and E. Ghani. "How Significant are Externalities for Development?" *World Development* 9:6 (1991): 569-594.

Wilson, Steven R. "Continuous Improvement and the New Competition: The Case of U.S., European, and Japanese Firms in the Mexican Maquiladora Industry." Unpublished Doctoral Dissertation, The University of Tennessee, Knoxville, TN, 1992).

Wood, Adrian. *North-South Trade, Employment and Inequality: Changing Fortunes in a Skill-Driven World*. New York: Clarendon Press, 1994.

Young, Alwyn. The Tyranny of Numbers: Confronting the Statistical Realities of the East Asian Growth Experience. *Quarterly Journal of Economics*. 110:3 (August 1995): 641-680.

Yotopoulos, P.A. and J. B. Nugent. *Economics of Development: Empirical Investigations*. New York: Harper and Row, 1976.

5

THE FIRM: A STRATEGICALLY LEARNING INSTITUTION

John Groenewegen
Erasmus University, Rotterdam

William Cole is an economist who is first of all interested in the dynamics of real institutions; he has shown a special interest in the strategy of Japanese firms. In this chapter I will try to characterize the Japanese firm as a dynamic institution in which routines, competence and strategy are central. This perspective on the firm is rooted in Penrose (1959), and Richardson (1972) and more recent dynamic approaches towards the firm which can be found in Groenewegen (1996). In order to be able to draw a sharp picture of the firm as an evolving, learning and strategic institution I start with a presentation of the more traditional theories of the nature of the firm. The traditional approaches belong to the neoclassical research program which is discussed in the first section. Then transaction cost economics is presented as a "go-between": on the one hand the neoclassical research program is left and the firm is not considered to be a production function or a nexus of contracts anymore, but the firm is analyzed as an institution. On the other hand transaction cost economics retains many characteristics of the neoclassical approach and most of all it is of a comparative static nature. Using the insights of transaction cost economics we discuss in the third section the building blocks of a dynamic theory of the firm, in which strategy, routines and path dependency become essential characteristics. After having discussed some methodological issues related to the application of different theories to a concrete case we turn our attention to the Japanese firm in the last part of this chapter.

TRADITIONAL THEORIES OF THE FIRM

With traditional theories of the firm we mean theories belonging to the neoclassical research program. Latsis (1976) characterized this program as one of "single-exit-situations." In the hard core of the theory the actors are decision-makers who have correct knowledge of their situation, who follow the rule of maximizing or minimizing behavior and because of this the actors act "appropriately to the logic of their situation" (Latsis 1976, 22). The normative part of the hard core, the heuristic, can be characterized as follows:

- construct static models; given for example the technology what then is the optimal size of the firm?
- minimize, or if possible, eliminate the non-economic content of the model; make the actor psychologically as empty as possible and reduce the relation with his physical and social environment to the minimum
- construct the situation of the actor so that the "logic of the situation" determines the outcome; the actors cannot logically choose something else than one optimal equilibrium solution. This is the single-exit-situation.
- the following rules in modeling should be applied: construct functions which are suitable for the application of procedures of calculus; if no determinate equilibrium results from the model then modify the situational assumptions such that one exit results; if the model yields a determinate equilibrium then introduce more realistic situational assumptions.

Latsis clearly shows how the neoclassical research program deals with multi-exit-situations in oligopolies, monopolistic competition and situations of limit pricing. Inherently multi-exit-situations are reduced to single-exit-situations: the factors that determine the equilibrium lie outside the decision-making units and the problem of interdependence among producers is assumed away. An example is the case of limit pricing as presented in Sylos's model:

> "Existing producers are assumed to act collusively. Potential entrants are assumed to be attracted to entry by the possibility of profits. However, the "possibility of profits" as seen by potential entrants depends on their expectation about the reaction of the existing producers to their entry" (Latsis 1976,35).

And he continues:

> "(....) potential entrants assume that existing producers will adopt *the most unfavourable strategy open to them with respect to a new entrant, namely that of maintaining output"* (Latsis 1976,36 italics in the original).

The neoclassical research program results in an empty, colorless picture of the firm: decision-makers without decision procedures do not act but only react. They do not compete in the sense of rivalry, but only react according to the way they are programmed.

The neoclassical theory of the firm, or better the theory of production functions, clearly has advantages. Because of the logic of the situation no attention in the theory has to be paid to the decision-makers' inner environment. Whether the actor is of an optimistic or pessimistic nature, what kind of information-gathering rules he follows, the decision rules he has to take into account, his learning procedures, all these psychological and social-psychological items do not have to be included the model. Because of that a mono-disciplinary approach is possible, which makes life much easier, especially when single-exit-situations allow for a "rigorous analysis." These advantages are so attractive that the research program of situational determinism has conquered a dominant position in the modern micro theory in the normative principal-agent theory as well as in modern developments in industrial organization known as the "New Industrial Organization" to which we turn our attention now.

AGENCY THEORY

Normative principal-agent theory is of the "mechanistic design"-type, in which agents can specify ex ante monitoring and incentive devices in such a way that violation of contracts will be minimized. The actors are neoclassical maximizers using opportunism to realize their goals. The question in agency theory is what the optimal contract looks like in a world with actors with different objectives who are hyperrational and of whom the attitude towards risks is given. Agency theory builds complex models with time lags, asymmetric information and the like, but its neoclassical nature is preserved: utility maximizing individuals operate in worlds in which the "logic of the situation" determines the optimal contract.

The firm is considered to be a nexus of contracts: in a perfectly symmetrical relationship employers (principals) negotiate with employees (agents) about the contracts, in which the transaction of labor is coordinated. It is explained how in team production the monitor has to prevent agents from shirking and how competition among monitors will prevent them from exploiting agents. Actors are assumed to be fully informed about prices, marginal productivities and the like and they are able to exit and renegotiate until equilibria result.

NEW INDUSTRIAL ORGANIZATION (NIO)

Shepherd (1990) discusses three recent developments in the school of NIO: the Chicago-UCLA school, the contestable markets school and the modern game theory of oligopolies. In all these approaches the neoclassical maximizing actor is so well informed that the situation of competition in the market, the potential competition or the structure of the game determine the optimal single exit. For instance in the Chicago-UCLA school it is assumed that competition selects the most efficient firm so the one which survives has necessarily minimized costs. Their efficient market hypothesis is not tested but presented as an a priori. A point of view with far reaching implications for competition policy: monopolies in the real world are efficient because they survived the competition battle so let them alone.

In contestable market theory the emphasis is not on existing market structures and competition, but on potential competition. In most cases monopolists will behave as if there was competition among many because they feel the pressure of potential competition. In contestable market theory the world is dominated by situations of "hit and run": free entry and free exit for firms not hindered by sunk costs.

In game theory the logic of the situation is not determined by existing or potential competition, but by the rules of the game. Schotter (1994) characterizes game theory as follows:

- game theory allows us to describe and analyze social and economic situations as if they were a game of strategy
- a game of strategy is an abstract set of rules that constrains the behavior of players and defines outcomes on the basis of actions taken by the players
- the rules specify:
  a?? who the players are
  b?? the order in which the players will make their moves and the choices that will be available to the players as well as the information the players will have when they make their moves
  c?? how much utility each player will receive depending on the choices of all players in the game

In other words: in game theory the heuristics informs the researcher precisely how to construct one-exit-situations.

In sum: traditional theories of the firm are of a methodologically individualistic nature and the analysis is comparative static. Given the technology, what then is the optimal size of the firm? If technology, the risk attitude, the potential competition and so on change, what then is the optimal solution? The analysis is such that theorists first aim at the construction of single-exit-situations.

TRANSACTION COST ECONOMICS (TCE): THE "GO-BETWEEN"

Oliver Williamson developed the idea of transaction costs as formulated in Coase (1937). After his *Markets and Hierarchies* (1995), Williamson developed four categories of governance structures (market-, bi-, tri- and unified governance), which he matched with categories of transactions (degree of asset specificity, frequency and uncertainty). Different types of contracts (classical, neoclassical and relational) are linked with the governance structures. Actors are bounded in their rationality and can behave opportunistically. Actors are assumed to minimize production and transaction costs. In doing so ex post the most efficient governance structure is supposed to survive. Understanding the institution of the efficient firm is the main issue in TCE, although Williamson admits that explanatory variables other than efficiency can play a role (especially in the organization of work). It is his strong conviction that economic efficiency is the most important explanatory variable for understanding and explaining firms. A selection process is assumed to be at work assuring care of the survival of the fitter. Williamson presents his TCE as a comparative institutional analysis. After a description of the characteristics of the transaction (asset specificity, opportunism and the like), Williamson shows the difficulties in drafting and monitoring market contracts which offers an opportunity for more efficient governance structures like firms of different structures (U-form, M-form, conglomerate, transnational corporations and the like). Governance structures exist because others fail due to relatively high production and transaction costs. This approach is static: given the technology, given the asset specificity, given opportunism and given the other variables in the "organizational failures framework", which organizational structure is most efficient? The question how governance structures develop over time is not at stake. Apart from the so-called "fundamental transition" of a market with a large number of suppliers to a situation of small numbers, dynamics is no part of TCE. The main causalities are shown in figure 5.1 with the solid arrows.

Figure 5.1. A Layer scheme

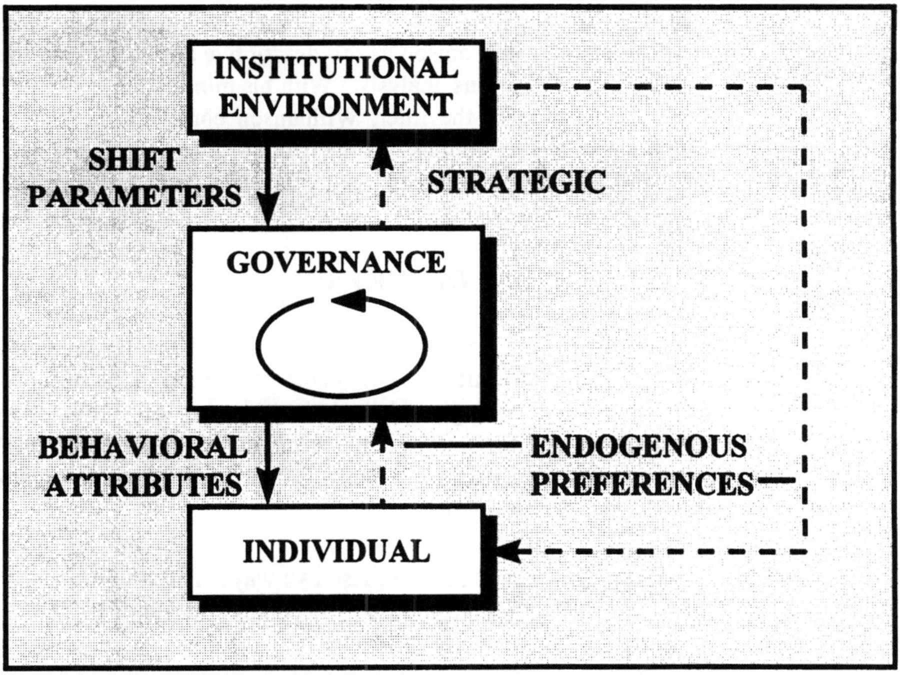

Source: Williamson

With respect to the institutional environment Williamson follows the distinction of North and Thomas (1971) between institutional environment (political, social and legal ground rules, also called the "belief system") and institutional arrangements (that govern the way in which units can cooperate and/or compete). The institutional environment is considered important by Williamson, and changes in that environment alter the comparative costs of markets, hybrids and hierarchies. Critics explain that such a recognition is not sufficient and that a detailed understanding of the institutional environment should always be the starting point of research of economic organization. An analysis of, for instance, the subcontracting relations of a firm should start with a detailed study of the institutional environment i.e., the legal rules (contract law, competition law), the political pressures (ministries stimulating network relations), the social attitude towards cooperative business relations, the cultural-historical development and the like. When these conditions are clear it is possible to understand why firms in Japan, for instance, chose subcontracting as an efficient governance structure and why firms in the US opted more for vertical integration.

The other solid arrow in figure 5.1 runs from the individual level to the governance structure: given the characteristics of bounded rationality and opportunism, individuals will choose the cost minimizing structure. Other relations exist between the three levels, but these are considered by Williamson as of secondary nature (the dotted arrows). Williamson presented himself in his *Markets and Hierarchies* as a “New Institutionalist” expressing the fact that he was interested in explaining the firm not as a production function, but as an institution. He also wanted to make clear that he was different from the American institutionalists like Thorstein Veblen. Their work is considered by Williamson as too descriptive and of lacking the theoretical base necessary for a “rigorous analysis.” With his minimizing actor, and his assumption about the selection of the fitter, Williamson considers the economizing lens the main issue in the economics of organization. Because of this neoclassical nature, but with the opening to institutional analysis, Oliver Williamson’s work is considered here to be a “go-between.”

TOWARDS A DYNAMIC THEORY OF THE FIRM

The question about the nature of the firm was raised by Coase (1937). He indicated that the “ordinary technique of economic analysis” could be applied to the question of economic organization. After having explained the concepts of combination (transactions previously organized by two entrepreneurs become organized by one) and of integration (transactions previously organized in the market become organized within the firm), Coase (1937,81) writes:

> A firm can expand in either or both of these two ways. The whole of the “structure of competitive industry” becomes tractable by ordinary technique of economic analysis.

As explained above Oliver Williamson further developed that economic analysis of the firm. The question about the nature of the firm was also raised but then from a different angle in Penrose (1959) and Richardson (1972). They also picked up Coase's question, but tried to find the answer along another route: capabilities and competence as a firm's specific identity became central and they laid the ground for the modern evolutionary theory of the firm (Hodgson 1996). Like Coase, Richardson raised the question of why transactions are coordinated inside organizations, via markets or through relations of cooperation. For Richardson the basic unit of analysis is the "activity" of the firm, which can be similar or dissimilar. Similar activities demand the same capabilities, that is to say the same knowledge, experience and skills. Penrose (1959) explains the growth of firms also in terms of capabilities and explains how these capabilities develop. Nowadays we would talk about "core activities" and "path dependency." Richardson also introduces the concept of complementarity: activities are complementary when they represent different phases of a process of production and require coordination. Having introduced "activity, capability and complementarity", Richardson argues that similar activities can be efficiently organized in a form called "direction": activities are then subject to a single form of control and fit into one coherent plan. For dissimilar, complementary activities the market is an efficient organization because aggregate demand and supply are matched ex post and stocks and prices are indicators for producers and customers to adapt their plans. Richardson introduces a third type of transaction in order to explain the coordination through cooperation between firms. If activities are closely linked like a cigarette with an appropriate filter tip, then

> (...) we require to match not the aggregate output of a general-purpose input with the aggregate output for which it is needed, but of particular activities which, for want of a better word, we might call *closely complementary* (Richardson 1972, 891, italics in the original).

In case of dissimilar, closely complementary activities parties need to coordinate their plans ex ante for which some form of cooperation is suitable.

With the introduction of type of activities and type of coordination mechanisms Richardson can be seen as an evolutionary economist "avant la lettre." With the introduction of the concept of capabilities he extends the work of Penrose and lays the ground for the modern theories of the firm which incorporate ideas of competence, learning and path dependency. From that perspective the nature of the firm is the cumulative causal process in which capabilities are accumulated. The existing capabilities more or less determine the path along which the firm can develop.

With respect to transactions the point should be made that transactions should not be analyzed in isolation, but in the context of their interdependencies. Transactions form a coherent whole tied together by a common knowledge base. The asset of a firm is this knowledge base with a supporting system of norms, rules and structures. The growth of the firm has now become endogenous: the learning capability of the firm determines its growth. It is very well possible that for firms a new technology is available, but because of the knowledge base and the "mental maps" the firm simply does not "see" that opportunity. On the other hand, firms develop their own teams with specific knowledge and can develop technologies and new products that other firms cannot. Important in this institutional-evolutionary approach is the protection the firm offers to the individual participants. In TCE the participants become vulnerable towards opportunism in case of asset specificity which calls for contingency contracts or special governance structures; in the institutional approach firms are political institutions which organize the cooperation among the members and protect individuals against rent-seeking behavior of others. Through all kind of norms, rules and procedures, the exploitation of the capabilities of individuals by others is limited. The nature of the firm has now become its organization with values, norms, rules and structure as safeguard, which make specific investments possible.

Leaving the comparative static world of neoclassical theory and transaction costs economics, firms turn out to be both subject to change and to initiate change. If this issue is considered important then theories of the firm should be appropriate for analyzing the dynamics of the firm. What this means for the extension of the theoretical framework can be explained in terms of the layer schema in figure 5.1. We first have a look at the level of the individual.

With respect to the individual Williamson considers bounded rationality and opportunism crucial. In several contributions (John 1984; Noorderhaven 1996) the need of endogenization of opportunism and the importance of trust are discussed. The "homo economicus" should be replaced by a "homo socialis", whose behavior can only be properly understood in relation to his environment. Opportunism can be relevant depending on the conditions, therefore not only the degree of asset specificity is important, but a number of conditions like attitudinal orientation, bases of power (coercive or legitimate), and type of bureaucratic organization (see John 1984) play a role in understanding opportunism. Other conditions are important for understanding trust (see Noorderhaven 1994) like consistency, discretion, loyalty, openness, receptivity and the like.

In terms of figure 5.1: the dotted arrow from the institutional environment to the individual becomes a solid one, which has large implications for the understanding of individual behavior. Now it becomes understandable why transactions with the same "amount" of asset specificity do not invoke opportunistic behavior in, for example, Japan but do so in the U.S. The preference function of the Japanese entrepreneur influenced by his institutional environment differs with respect to opportunistic behavior. For the governance structure this implies that the same transaction is coordinated in Japan via, for instance, a relation of subcontracting, whereas the same transaction is internalized in the U.S. through vertical integration. This should not be only understood in a static sense of the institutional environment being a shift parameter. On the contrary: what a dynamic theory of the firm should offer is an insight into the process which develops between the institutional structure, the individual and the governance structure in which all of them are interdependent and re-enforcing. So the extension of TCE into the direction of that interdependency is necessary to understand better the relation between the (social) individual and the governance structure. But there is more.

In recent discussions the issue of epistemology is also raised (Nooteboom 1993 1996), which should have far reaching implications for the theory of the firm. In neoclassical as well as new institutional theories, the firm is considered to be something objective to be "discovered" if actors just put enough time, energy and costs in their search. The underlying reality of, for instance, technology or preferences can be revealed if sufficient information-, search-, storing-, and evaluation costs are made. This epistemology, which is implicit in the traditional theories of the firm, is called "metaphysical realism." This approach is questioned among others by Nooteboom (1993) and Lawson (1995). They depart from the position that reality is partly structured by the concepts and theories researchers have in their "inner environment." In interaction with their physical and social environment, actors on the one hand "construct" reality with a specific theoretical framework, whereas on the other hand the interaction with other researchers and the physical environment influences the adaptation of the theoretical framework. In other words: people's inner environment is filled with so-called categories which are used to structure reality. In doing so people interact which each other; a communication which is only possible if the same categories are applied.

People also interact with their environment and from the experiences their categories develop. What is an adequate category to analyze, for instance, a firm is a matter of communication with other researchers (intersubjectivity) as well as a matter of interaction with one's physical environment out of which it becomes clear how explanatory the theory is. This epistemology is called "social constructivism" (Nooteboom) or "critical realism" (Lawson). Cognition now becomes a matter of learning and path-dependency. With this other epistemology of social constructivism the concept of bounded rationality also gets another meaning; the boundedness is not only a matter of limited computational capacity, but also a matter of being equipped with specific categories which limit the capacity to "see" other "parts" of reality. The idea of social constructivism corresponds closely with recent work of Denzau and North (see Groenewegen et al. 1995). Denzau and North (1994) point to the idea that individuals operate with specific "mental maps." The maps facilitate communication if individuals have the same map, but constrain actions for which other maps are required. Here the concept of proximity is helpful: individuals and firms can learn if there is enough proximity with the mental maps of others; if the distance between the maps is too far, no contact, communication and learning is possible (see especially Nooteboom 1996 on this).

The application of the epistemology of social constructivism has large implications for the theory of the firm. Reality is no longer to be known objectively by firms; it is very well possible that some efficient governance structure is not chosen simply because it does not fit into the mental maps of the actors. With this epistemology it becomes possible to understand why, for instance, Japanese firms select different governance structures for the same type of transactions than American firms, or why such differences are shown between industries, or even within the same industry between firms with a different path of development.

We now turn attention to the arrow in figure 5.1 which runs from governance structure to institutional environment.

THE STRATEGY OF THE FIRM

From the meso level of the governance structure the arrow of strategy indicates an influence of the firm on the institutional structure. In order to develop its knowledge base and to realize its "mission" the firm also tries to influence the institutional structure. Clearly the basic values and norms are more constraints for actors than strategic variables, but the legal rules and the socio-political structures can be subject to strategic decision-making. Now changing institutions is not something firms easily can and will undertake, because this demands fundamental changes in the mental maps, existing routines, etc. Only when the firm is under severe pressure for a long time, is management pushed to look for other strategies which probably demand changes in the institutional structure as well. The different steps involved in the change of strategy and institutional structure are shown in figure 5.2.

Figure 5.2.

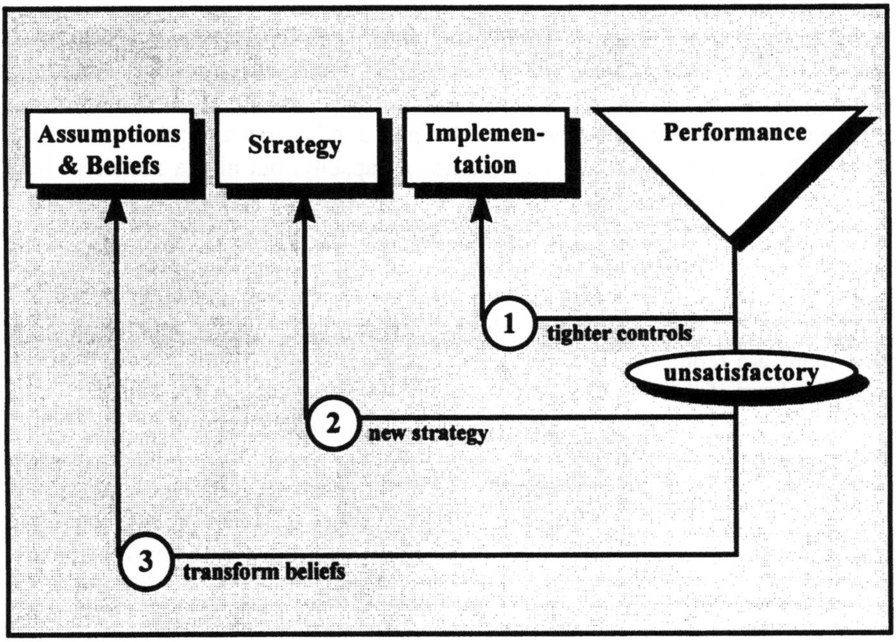

Source: Dietrich 1994

The first reaction of management towards an unsatisfactory performance of the firm is to strengthen the existing policy: more of the same. When this is not working management is pressed to look for other strategies. For instance, firms which are used to following the strategy of doing their own R & D cannot easily switch to a strategy of cooperation in strategic alliances. Only when a continuation of the existing strategy does not result into the desired performances is management forced to look into such new worlds. Whether they are able to do that largely depends on their mental maps, on their cognitive capabilities which are influenced by the interaction with other firms in their proximity. For instance, firms which are used to working in networks with close relations with subcontractors, or in joint ventures with competitors, are in a better position to change strategy than firms which operate in isolation. The question is, do firms know about the experiences with other strategies? Do they communicate with managers who followed other strategies based on other mental maps? If not, then a change of strategy is problematic. Now if strategy changes at a large scale (for instance a sector wide change from vertical integration to subcontracting), then pressure of economic actors to change the institutional structure can be such that they are politically successful (for instance a change in the competition law towards a more positive attitude with respect to agreements among competitors). In terms of figure 5.1 the above discussion should be shown by a change of the dotted arrow between governance structure and institutional structure into a solid one.

With respect to the governance structure having a life of its own, Williamson recognizes such effects and makes the point that "such effects demonstrates the need for deep knowledge about organizations, but it does not imply that the economic approach to organization (which easily misses such effects) is fatally flawed" (Williamson 1993, 11-12). Indeed critics point to the need of a "deep knowledge about organizations" and explain that corporate cultures can be so specific that governance structures that should exist according to TCE-insights are not selected simply because such transaction cost minimizing governance structures are not part of the mental map of the firm. The argument of social constructivism and path dependency also applies to corporate culture.

METHODOLOGICAL REMARKS

Before we try to understand the Japanese firm by applying the theoretical insights discussed above it seems useful to make a few methodological remarks (see also Groenewegen and Vromen 1996). In our point of view the theories discussed mostly refer to different issues. In other words the problem definitions for which the theories are developed are different, which implies that the theories are complementary and not contradictory. The traditional theories, and also transaction cost economics, are first of all constructed for solving static questions of optimal solutions, whereas the institutional theories have change, procedures of decision-making, development of strategies, learning and path dependency as their core questions. In our point of view none of the theories can claim universality (yet) and it seems to be adequate to consider the different theories as complements. In short, different, theories can live happily next to each other as "different theories relevant for different issues." Now relevancy of theories does not only depend on the issues at hand, but also the prevailing conditions in the situation under study determine whether a theory is relevant. When, for instance, in a theory selection of the fitter is assumed to take place because of severe competition and the market structure excludes competition, then that specific theory is not relevant for the specific situation at hand. In short, "different theories are then relevant under different sets of conditions." Determining the issues and conditions which make the one theory more relevant than the other in a concrete case of research is here considered to be fundamental.

THE JAPANESE FIRM

In this section we will try to understand the structure and strategy of the Japanese firm using the theoretical insights discussed above. We depart from transaction cost economics as the "go-between." Our understanding of the Japanese firm is supposed to improve as we discuss the layers and arrows of figure 5.1.

THE FIRST LAYER: THE JAPANESE INSTITUTIONAL ENVIRONMENT

The institutional structure is defined as the system of norms, legal rules and structures (market as well as socio-political ones) within which the actors perform at meso and micro levels. So for understanding the structure and strategy of Japanese firms, knowledge of the institutional structure is necessary. With respect to the system of values, anthropologists often refer to Confucian values: Hofstede and Bond (1988) for example come to the conclusion that next to the elements of the acceptance of power, of the collectivist instead of individualistic values, of the importance of masculine instead of feminine values, the characteristic of Confucian dynamism is of crucial importance in the Japanese structure. In short, the values result in norms of behavior which favor group-orientedness (collectivism) and adherence to one's role (stability in relationships, see Orru, Hamilton and Suzuki 1989), as well as a positive attitude towards dynamics (incremental change, see below). These basic values and norms are reflected in the role of government in relation to firms, in the relation of the bank toward the members of the same industrial group, in the relation between a parent firm and its subcontractors, etc. With respect to the legal rules which are based on the value system researchers explain that these favor the existence of long-term relationships, of cooperation in groups, and of collective decision-making (Imai and Itami 1984). A good example is Japan's competition law which opens ample opportunity for cooperation among competitors and makes even cartels possible in depressed sectors.

With respect to Japanese socio-political structures, these also reflect basic values: close relations between politics, bureaucracies and management of large enterprises form "old boys networks" (Johnson 1982) which, to a large extent, influence the costs of transactions.

THE SECOND LAYER: THE JAPANESE GOVERNANCE STRUCTURE

The governance structure that is typical for Japan is the firm as part of industrial groups. The Japanese firm operates in a specific environment which fosters the development of long-term relations. The specific characteristics of, for instance, the internal labor market, the role of the bank, and the influence of government are all interrelated. We start with a description of the situation inside the individual firm and will proceed with the external relations within the industrial group.

THE FIRM

Typical of the internal structure is the processing of information inside the firm and the incentive system related to the labor market (Aoki 1990; Odagiri 1994). Information is produced at all levels in the organization and each level has direct contact with customers. Lower operational levels have specific information and handle problems autonomously. Employees are supposed to invest in specific skills, to delegate responsibility to lower levels and to learn the next generation skills specific to the job. Our "go between" of transaction costs economics would point to problems of opportunism here, but Williamson (1991) explains that typical Japanese institutions act as safeguards against opportunism. In the Japanese firm, opportunism is attenuated because of lifetime employment contracts and slow internal promotion based on a range of criteria like skills, experience, cooperativeness and the like. Intergenerational teaching, delegation of responsibility and specific investments are rewarded with a higher ranking, whereas the slow, well managed promotion system hinders young colleagues from bypassing older persons with knowledge that they have taught them. In this the personnel department and the enterprise union play an important role. The personnel department is an organizational innovation taking care of the careers in a comprehensive, long-term way, so the individuals are protected from short-term actions of superiors. Confidence that the specific investments they make will be beneficial for their own career is bolstered by the protection from the personnel department. The enterprise union explicitly operates in line with the objectives of the enterprise (not the industry) and offers an opportunity ("voice") to the employees opposite the personnel department. From the economizing point of view the personnel department and the enterprise union are governance structures facilitating a "deepening and more effective deployment of firm-specific capital" (Williamson 1991,88).

In Japan an imperfect external labor market (difficult to find a comparable rank by another firm) is necessary to prevent well-skilled employees from being bought away by competing firms. The firm itself does not recruit personnel at middle and higher level in order not to distort career perspectives of junior employees (Odagiri 1994). So information processing, internal decision making, an imperfect labor market and ranking are all interrelated. A deeper understanding of these interdependencies can arise in applying concepts of the dynamic theory of the firm. The internal labor market also fosters the harmonization of the mental maps of the individuals. Job rotation and the process of information exchange and decision-making bring individuals in contact with varieties close to their own mental maps so a process of learning is possible. Management purposefully protects these institutions because these are crucial for realizing the firm's mission. The elements discussed above form a system which cannot be fully understood in terms of minimizing production and transaction costs. The concept of the firm as a political coalition with its own knowledge base and own path of development is indispensable for understanding the internal labor markets and its supporting institutions. Also insights in the external relations which are closely related to the internal structure, is necessary.

To that we turn our attention in discussing subcontracting, the role of the bank, and the sogo shosha.

SUBCONTRACTING

Because of the internal labor market the Japanese firm is "capability-driven" (Imai and Itami 1984) resulting in a workforce with similar skills, which is favorable for information processing and learning. Japan has many so-called production keiretsu ("sangyo keiretsu"), where a parent firm has economic (supplier-buyer), financial (stockholding), organizational (management assistance), technological (transfer of technology) and personal relations with subcontractors, co-designers and co-makers. With the first tiers of subcontractors the relations are long term and subcontractors depend heavily on the parent. Our "go between" of TCE points to opportunism because of the specific investments made by the subcontractors. However, through specific hybrid modes of organization, the incentives for opportunistic behavior are considerably reduced. In the economizing approach attention is focused on the ranking of subcontractors in categories A through D (Asannma 1989). The relation with categories A and B are of specific importance and carefully nurtured by the parent firm; the C category is the buffer, whereas D is going to be eliminated. The relations between the parent and the different tiers of subcontractors are governed in relational contracts which fit into the wider institutional environment. Bilateral monopolies with all the risk of opportunism are avoided: also in categories A and B there is competition in situations of contract renewal, while parent firms mostly create alternative sources of supply (for instance double sourcing). According to the economizing point of view subcontracting is first of all an organizational devise to allocate resources efficiently and to offer the core firm the possibility to specialize in specific human capital. The subcontracting system is interrelated with the internal organization of the Japanese firm: because of the tiers of subcontracting the parent can concentrate on the activities which demand specific capabilities and can implement a ranking and lifetime employment system.

From a dynamic point of view subcontracting is to be understood as a method to open the door to close but different mental maps. Through close relations with the co-makers and co-designers of the first tier parent firms enlarge their cognitive capabilities. Recent developments also show horizontal cooperation between subcontractors which clearly stimulates the process of learning about new developments in demand conditions and technology. These developments are characterized as INTAC (Inter-industrial Network for Technological Activities, see Furukawa, Teramoto and Kanda 1990), or "Igyoushu Kouryu" (Van Kooij 1991). According to these developments the traditional pyramid of subcontracting relations is replaced by a cobweb. These more horizontal cobwebs stimulate learning among subcontractors which can also be beneficial for the parent firm. The system of subcontracting clearly can be understood in terms of efficiency. However, why the system developed in Japan and how the relationship is governed can only be properly understood when developments in relations with the institutional environment and other institutional arrangements are taken into account.

THE MAIN BANK

In the "kinyu keiretsu" (horizontally structured group), industrial firms, the general trading company (GTC) and the group bank (main bank) are connected through economic, organizational and financial relations. The firms inside the group have bank accounts with the group bank, the group bank is their major stockholder (cross stockholding), the group bank is the most important lender to the group members, and the group bank is the one that monitors the group members and if necessary organizes a reconstruction. The group bank is important, but that is not to imply a complete dependency by the industrial firms; group members are free to develop relations with other banks, to have stockholdings with other banks and when the firm is doing well the bank does not interfere in the firms strategies. However, if a group member is in trouble, the group bank is responsible for restructuring, rationalization and replacement of management. Firms are ranked by the group bank based on the financial position and duration of the relationship. Highly ranked firms are very independent and with their stable shareholders like the group bank, these firms do not have to fear hostile takeovers.

Our "go between" of TCE explains the role of the group bank in terms of economizing. The group bank is responsible for the group members and has a reputation to defend as an efficient monitor and lender of last resort. The high ex post cost of reconstruction is an incentive for close intensive ex ante monitoring. From an economizing point of view the business group and the role of the main bank is an example of collective organization where reputation effects can usually be expected to foster reliability (Goto 1982; Aoki 1990, and Williamson 1991). In order to understand these reputation effects a deeper knowledge of the Japanese institutional structure and, especially, of the norms and structures that guide behavior is necessary. Among others (Orru, Hamilton and Suzuki 1989) point out that the bank in the "keiretsu" plays an intermediate role as an agent on behalf of all group members. This role is enforced by the administrative guidance of the Ministry of Finance which expects main banks to play a monitoring role. If not, the ministry will take disciplinary measures such as a stop in permission to open new branches. Insights from the dynamic theory of the firm are furthermore relevant to understand the role of the bank in the learning process of group members. The role of the main bank can also be understood in terms of the dynamics of capabilities and learning: through the meetings in the President Club (Goto 1982; Orru, Hamilton and Suzuki 1989) the banks learns a lot about the mental maps of group members. In the process of information exchange these mental maps become known to the members of the group. In case of reconstruction and change of strategy the bank is in a position to inform firms of possibilities they cannot think of themselves. Also in case of reconstruction the bank is in position to replace management so other strategies can be implemented.

THE GENERAL TRADING COMPANY

Traditionally the general trading company (GTC or "Shogo Shosha") fulfills several functions for group members. Traditional neoclassical theory points to the GTC as realizing economies of scale and scope in producing information and organizing trade. Traditional theory also point to the reduction of risks with respect to the fluctuations in foreign currencies (the GTC imports and exports from the same country) and with respect to variations in demand (the GTC trades in many different products so a decline can be compensated with a rise elsewhere). Our "go between" of TCE explains the GTC in terms of asset specificity, opportunism and safeguards (Roehl 1983; Shin 1989). Suppose a trading company takes care of the export of a product with very specific characteristics. Suppose further that the company has undertaken specific investments in order to produce information about consumers and has also made the product known to the public by means of an intensive advertising campaign. Then all the conditions for opportunism of the manufacturer are fulfilled: he can bypass the GTC and make opportunistic use of the prior investments made by the trading company. Shin (1989) explains that the GTC has possibilities for safeguarding its investments, but because of rising transaction costs, it can never safeguard itself completely in cases of asset specificity. TCE concludes that GTC are efficient in cases where transaction specific investments are low and possibilities for opportunism are small. In case of specific investments, internalization of the trade inside the manufacturing firm will be efficient. Facts confirm this hypothesis: trading companies have a relatively large share of trading in commodities like food and tobacco, textiles and wood, pulp and paper. Their role is smaller in categories like machinery and equipment. The analysis of Shin (1989) also shows that the role of trading companies is larger in import than export, which is understandable because Japanese exports are more sophisticated.

Complementary to this the dynamic theory of the firm points to changing strategies and learning. This is relevant for the Japanese case because the GTCs have shown unsatisfactory results for some years now. Because the exports of Japan become more and more specific, the GTCs loose market share to the manufacturing firms which take care of the exports themselves. The change in strategy GTCs are forced into is one of specialization: instead of being "general" the trading company should become specialized in, for instance, machinery, consumer electronics, or fine chemicals. Being a member of the "keiretsu" means that the expertise to specialize is present and under the umbrella of the bank and leading industrial firms, such a change in strategy by the trading companies is possible.

THE JAPANESE FIRM HAVING A LIFE ON ITS OWN

Over time firms develop their own corporate culture, their own routines and decision-making procedures. With its own identity the firm becomes a political institution, a "going concern" that follows more or less its own route of development. Being "capability driven" the diversification strategy of the Japanese firm is a related one: the firm in "having a life of its own" will try to make optimal use of its capabilities. Related diversification is reinforced by the internal decision making process (in a bottom-up process it is not likely that unrelated radical diversification will be proposed). Also the role of banks which finance the diversification strategy will reinforce related diversification (financial institutions prefer businesses to stick to their core activities which is less risky). Related to this is the strategy of innovation: Japanese firms with their internal labor market perform well with respect to incremental innovations and less well with respect to radical innovations. So a deeper understanding of the Japanese firm as "a life on its own" demands the application of the theoretical insights of the dynamic theory of the firm to issues of resources-based learning and path dependency based on capabilities.

CONCLUSION

In this chapter we have tried to offer a theoretical framework for understanding the modern firm in its environment. We have discussed and applied traditional theories of the firm, the "go between" of TCE and the dynamic theory of the firm. We have argued that the different theories can be considered complementary and that their relevancy depends on the issues and conditions at hand. In order to improve the theoretical framework and to be able to be more explicit about "issues and conditions" comparative empirical research of the dynamics of firms in different institutional settings seems useful.

REFERENCES

Aoki, M. "Towards an Economic Model of the Firm." *Journal of Economic Literature* 4:28 (1990): 1-27.

Asanuma, B. "Manufacturer-Supplier Relationships in Japan and the Concept of Relation-Specific Skill." *Journal of the Japanese and International Economies* 3 (1989): 1-30

Coase, Ronald. "The Nature of the Firm" in Louis Putterman, ed. *The Economic Nature of the Firm: A Reader*, pp. 72-86. Cambridge: University Press, 1993.

Denzau, A.T. and D. C. North. "Shared Mental Models: Ideologies and Institutions", *Kyklos* 47 (1994): 3-31.

Dietrich, M. *Transaction Cost Economics and Beyond: Towards a New Economics of the Firm.* London: Routledge., 1994.

Furukawa, K., Y. Teramoto, and M. Kanda. "Network Organisation for Interfirm R&D Activities: Experience of Japanese Small Business." *Technology Management* 5 (1990): 418-438.

Goto, A.. "Business Groups in a Market Economy." *European Economic Review* 19 (1982): 53-70.

Groenewegen, J (ed.). The Japanese Group in The Dynamics of the Firm. Aldershot: Edward Elgar, 1993.

________ (ed.). *Transaction Cost Economics and Beyond.* Dordrecht/Boston: Kluwer Academic Publishers, 1996.

________, and J. J. Vromen. *A Case for Theoretical Pluralism in Transaction Cost Economics and Beyond.* In J. Groenewegen, 1996.

________, F. Kerstholt, and A. Nagelkerke. "On Integrating New and Old Institutionalism: Douglass North Building Bridges." *Journal of Economic Issues* 29:2. (June 1995): 467-475.

Hodgson. G. M.. *Corporate Culture and the Nature of the Firm in Transaction Cost Economics and Beyond* In J. Groenewegen, 1996.

Hofstede, G., and M. Bond. "The Confucius Connection: From Cultural Roots to Economic Growth." *Organizational Dynamics* (Spring 1988): 5-21.

Imai, K. and H. Itami. "Interpenetration of Organization and Market." *International Journal of Industrial Organization* 2 (1984): 285-310.

John, G. "An Empirical Investigation of Some Antecedents of Opportunism in a Marketing Channel." *Journal of Marketing Research* 21 (1984): 278-289.

Johnson, C. *MITI and the Japanese Miracle*. Stanford, California: Stanford University Press, 1982.

Kooij, E. van. "Japanese Subcontracting at a Crossroads." *Small Business Economics* 13 (1991): 145-154.

Latsis, S. J. *A Research Programme in Economics in Method and Appraisal in Economics*. In *Method and Appraisal in Economics*, J. S. Latsis, ed. Cambridge: Cambridge University Press, 1976.

Lawson, T. "Realist Perspective on `Economic Theory'", *Journal of Economic Issues* 29 (March 1995): 1-32.

Lindenberg, S. *Short-term Prevalence, Social Approval and the Governance of Employment Relations in Transaction Cost Economics and Beyond.* In J. Groenewegen, 1996.

Noorderhaven, N.G., *Opportunism and Trust in Transaction Cost Economics in Transaction Cost Economics and Beyond.* In J. Groenewegen, 1996.

Nooteboom, B. *Networks and Transactions: Do They Connect? in The Dynamics of the Firm*. In J. Groenewegen, 1993.

Nooteboom, B. *Towards a Learning Based Model of Transactions in Transaction Cost Economics and Beyond.* In J. Groenewegen, 1996.

North, D., and R. Thomas. *The Rise of the Western World: A New Economic History.* Cambridge: Cambridge University Press, 1973.

Odagiri, H. *Growth Through Competition, Competition Through Growth.* Oxford: Clarendon Press, 1994.

Orru, M., G. Hamilton, and M. Suzuki. "Patterns of Interfirm Control in Japanese Business." *Organizational Studies* 10 (1989): 549-574.

Penrose, E. T. *The Theory of the Growth of the Firm*. Oxford: Oxford University Press, 1959.

Richardson, G. B. "The Organisation of Industry." *Economic Journal* 69 (1972): 223-237.

Roehl, T. "A Transaction Cost Approach to International Trading Structure: The Case of Japanese General Trading Companies." *Hitotshubashi Journal of Economics* 24 (1983): 119-135.

Schotter, A. *Microeconomics, A Modern Approach.* New York: Harper Collins, 1994.

Shepherd, W.G. *The Economics of Industrial Organization.* Englewood Cliffs, NJ: Prentice-Hall, 1990.

Shin, K. S. "Information, Transaction Costs and the Organization of Distribution: the Case of Japan's General Trading Companies." *Journal of the Japanese and International Economics* 3 (1989): 292-307.

Williamson, O.E. "Strategizing, Economizing, and Economic Organization." *Strategic Management Journal* 12 (1991): 75-94.

________. *Markets and Hierarchies.* New York, NY: Free Press; 1995.

________. "Transaction Cost Economics and Organization Theory." *Industrial and Corporate Change* 2: 2 (1993): 107-156.

________. *Efficiency, Power, Authority and Economic Organization in Transaction Cost Economics and Beyond.* In J. Groenewegen, 1996.

6

AN OPPORTUNISTIC INTERPRETATION OF ADAM SMITH'S INQUIRY INTO THE NATURE AND INTERRELATIONS OF INVENTIONS, ENTREPRENEURSHIP AND INNOVATIONS

Hans E. Jensen
The University of Tennessee, Knoxville

According to the conventional wisdom in doctrinal history, "J. B. Say ... was the first [economist] to assign to the entrepreneur -- per se and distinct from the capitalist -- a definite position in the schema of the economic process." Adam "Smith glanced at the type occasionally," but he paid no attention to entrepreneurs beyond this furtive glance, Joseph A. Schumpeter alleged. Hence the "impression that his readers get" is that Smith was of the opinion that a "business runs by itself" (Schumpeter 1954, 555). Robert F. Hebert and Albert N. Link echoed Schumpeter when they observed that "Smith failed to separate the entrepreneurial decision making from among the various kinds of 'industrious people' in the economy" (Hebert and Link 1988, 46). The claim by Schumpeter *et al.* that Smith disregarded the entrepreneurial function has not gone unchallenged. Some forty years ago, "F. Redlich and B. F. Hoselitz ... showed the baselessness of such a claim." Recently, Enzo Pesciarelli demonstrated that the "concept of the entrepreneur can indeed be found in ... [Smith's] *Wealth of Nations*." He located three types of this personage therein: The "adventurer," the "projector," and the "undertaker" (Pesciarelli 1989, 521, 522, 523, 526).

In the following, I shall be concerned only with Smith's discussion of the role and activities of the "undertaker," Smith's term for "the only one" of the three types of entrepreneur that received his "sympathetic treatment and esteem" (Smith 1981, 181; Pesciarelli 1989, 526). Smith praised the undertaker because he viewed this economic actor as the promoter *par excellence* of the growth of those "necessaries and conveniences of life" that alone could cause an increase of the "happiness of mankind" which was the "purpose intended by the Author of nature, when he brought ... [its members] into existence" (Smith 1981, 10; and 1982a, 166).[1]

In the *Wealth of Nations*, the undertakers do not operate in a motivational vacuum, however. On the contrary, they strive to satisfy their desires for "gain" by acquiring new machines, tools and technologies from inventors who, like other "philosophers,"[2] engage in the act of invention "for its own sake, as an original pleasure or good in itself" (Smith 1981, 21; and 1982b, 51). In other words, inventors ply their trade in order to satisfy what Thorstein Veblen called their "idle curiosity" (Veblen 1957, 4).

In my discussion of entrepreneurial activities in the *Wealth of Nations*, I shall include in these pursuits certain types of performance that are not viewed conventionally as being entrepreneurial in nature. It is for this reason that the words "opportunistic interpretation" appear in the title of my inquiry. I use the term "opportunistic" in order to indicate that I have taken advantage of any opportunity that Smith gives me to interpret him as a creator of a paradigm in which improvements in the human condition are propelled by the interaction of inventors and innovating undertakers who engage in a multiplicity of entrepreneurial activities. If this interpretation is accepted, then Smith's work may be viewed as embodying in embryonic form some of the concerns -- and, perhaps, also some of the orientations and approaches -- of those authors in this volume who break new grounds in the related fields of (the new) industrial economics, Post-Schumpeterian evolutionary economics and Post-Ayresian institutionalism.[3]

In accordance with the above precis, this chapter is divided into four parts. First, I sketch the objectives that Smith wished to achieve through the publication of the *Wealth of Nations*. Next I summarize his inquiry into the nature and character of inventions. In the third part, I attempt to provide a fairly comprehensive interpretation of Smith's theory of innovation. In the conclusion, I speculate about the implications of Smith's work for those present-day scholars who labor in the vineyards of innovation and entrepreneurship.

SMITH'S CONCERNS AND OBJECTIVES

Smith wrote *An Inquiry into the Nature and Causes of the Wealth of Nations* with one overriding objective in mind: He wished to explain how rampant poverty among the "labouring poor" could be eliminated from this "great body of the people." Smith's principal measurement of poverty was the rate of "mortality ... among the children of the common people." He found, for example, that in "some places one half the children die before they are four years of age; in many places before they are seven; and in almost all places before they are nine to ten." Hence he concluded that it "is generally reckond [*sic*] that half of mankind die before 5 years of age. But this is the case only with the meaner and poorer sort" (Smith 1981, 99, 97; and 1982c, 193). Consequently, Smith's "sympathies went wholly to the labourer" (Schumpeter 1949, 353). As his first biographer put it, Smith wrote his *magnum opus* in order to contribute "to the happiness and the improvement of society" through the elimination of working-class poverty (Stewart 1794, 271).

As Smith saw it, it should, therefore, be society's goal to bring about a "universal opulence which extends itself to the lowest ranks of the people." This, he argued, could be accomplished only if the annual "produce" of the nation came to bear "a greater ... proportion to the number of those who are to consume it." Increase in output per hour of work would be the only means to this end. Smith had no doubt about the source of such a gain: "The greatest improvement in the productive powers of labour, and the greater part of the skill, dexterity, and judgment with which it is any where directed, or applied, seem to have been the effects of the division of labour" (Smith 1981, 22, 10, 13). He elaborated on this proposition as follows:

> This great increase of the quantity of work, which, in consequence of the division of labour, the same number of people are capable of performing, is owing to three different circumstances; first, to the increase of dexterity in every particular workman; secondly, to the saving of the time which is commonly lost in passing from one species of work to another; and lastly, to the invention of a great number of machines which facilitate and abridge labour, and enable one man to do the work of many (Smith 1981, 17).

Smith reversed the above order of the depicted processes when he analyzed their cause-and-effect relationships. Thus he argued that increases in output per hour of work are due to the substitution of machinery for labor in several operations.[4] Each worker is therefore burdened with fewer tasks and he becomes an efficient specialist in their execution. Moreover, in consequence of the reduction of the number of operations that he performs, each worker spends less time in moving from one task to another.

Smith's emphasis on the importance of the introduction of "proper machinery for executing [some of] the work" formerly performed by workers indicates that he was impressed by those contemporary economic changes that were wrought by nascent machine technology (Smith 1981, 104). In other words, he was fully "aware of the economic changes occurring during his lifetime, which coincided with the background and beginnings of the industrial revolution" (Hartwell 1976, 33-34).[5]

It is with this awareness of Smith as a backdrop that one has to interpret his writings on invention and innovation. Each of these activities, which he viewed as being performed by specialized agents, was described in some detail by Smith. Because inventors provided the hardware that undertakers innovated, I shall first turn to Smith's discussion of inventions.

SMITH ON INVENTION

Smith commenced his inquiry into the nature and processes of invention by observing that we "have not nor cannot have any complete history of the invention of machines, because most of them are at first imperfect, and [then] receive gradual improvements" from several individuals over long periods of time. He surmised, however, that in the misty past inventions were made by "those who [were to] use them." He speculated, for example, that it "probably was a farmer who made the original plow." Similarly some "miserable slave who had perhaps been employed for a long time in grinding corn between two stones probably first found out the method of supporting the upper stone by a spindle." Subsequently a "miln wright perhaps found out the way of turning the spindle with the hand" (Smith 1982c, 492).

> But [said Smith] he who contrived that the outer wheel should go by water was a philosopher, whose business it is to do nothing, but observe every thing. They must have extensive views of things, who as in this case bring in the assistance of new powers not formerly applied. ... [He] who first executed this, he must have been a philosopher (Smith 1982c, 492).

A comparable evolution occurred in manufacturing. "A great part of the machines made use of in ... manufactures ...were originally the inventions of common workmen." Inasmuch as the workers in question were "employed in some very simple operation, [they] naturally turned their thoughts towards finding out easier and readier methods of performing it" (Smith 1981, 20). When Smith began his search for the sources of economic growth and development, such work-related, or endogenous, inventions were being phased out in favor of inventions that occurred outside, or separate from, the work process. Smith described these exogenous inventions in the following manner:

> All the improvements in machinery, however, have by no means been the inventions of those who had occasion to use the machines. Many improvements have been made by the ingenuity of the makers of the machines, when to make them became the business of a peculiar trade; and some by that of those who are called philosophers or men of speculation, whose trade it is, not to do any thing, but to observe every thing; and who, upon that account, are often capable of combining together the powers of the most distant and dissimilar objects (Smith 1981, 21).

Thus like a twentieth-century American institutionalist, Smith argued that all inventions "result from the combination of previously existing tools ... and devices so as to form new tools ... and devices" (Ayres 1952, 56). The person who makes such a combination exogenously, whether "he was an artizan, or whatever he was.... must have been a philosopher," that is a technologist, engineer or scientist, Smith reiterated (Smith 1982c, 492). And he continued:

> When an artist makes any such discovery he shows himself to be not a meer [*sic*] artist but a real philosopher. ... It was a real phi-

> losopher only who could invent the fire engine [i. e. the steam engine], and first form the idea of producing so great an effect by a power in nature which had never before been thought of (Smith 1763, 570).

Undoubtedly, Smith was here referring to James Watt. "Smith had known Watt since 1757, when the latter was appointed mathematical instrument-maker to Glasgow University. It was in his workshop there that Watt repaired the Newcomen steam engine in 1764, leading to his own discoveries in that field of technology" (Mossner and Ross 1987, 248n). Thus it may be hypothesized that Smith formulated his concept of exogenous invention, at least partly, on the basis of conversations with Watt and his familiarity with the latter's inventive activities.

As Smith saw it, Watt and other makers of exogenous inventions were set apart from the rest of society by virtue of their "education" and the "habits" and "customs" that they acquired on the path leading to their emergence as inventive technicians (Smith 1981, 29, 152, 119). "Like every other trade," inventive philosophy "is subdivided into many different branches" so that "we have mechanical, chymical [*sic*] ... and physical" philosophy. This "subdivision of employment in philosophy ... improves dexterity and saves time. Each individual becomes more expert in his own peculiar branch, more work is done on the whole, and the quantity of science is considerably increased by it" (Smith 1763, 570; and 1981, 22).

It seems that Smith was of the opinion that there were two major causes of the demise of work-related inventions and the rise of exogenous inventions. One cause manifested itself as a decline of the inventive capabilities of workers and the other took the form of increased educational opportunities for a growing number of potentially inventive technicians. The following is Smith's description of the former phenomenon which, he argued, was one of the consequences of the increased division of labor that occurred in the wake of growing applications of machine technology in manufacturing.

> In the progress of the division of labour, the employment of the far greater part of those who live by labour, that is, of the great body of the people, comes to be confined to a few very simple operations; frequently to one or two. But the understandings of the greater part of men are necessarily formed by ordinary employments. The man whose whole life is spent in performing a few simple operations, of which the effects too are ... always the same, or very nearly the same, has no occasion to exert his understanding, or to *exercise his invention in finding out expedients for removing difficulties* which never occur. He naturally loses, therefore, the habit of such exertion, and generally becomes as stupid and ignorant as it is possible for a human creature to become... . His dexterity at his own particular trade, seems, in this manner, to be acquired at the expense of his intellectual ... virtues (Smith 1981, 781-782; emphasis added).

Moreover, "the division of labour, having reduced all trades to very simple operations, affords an opportunity of employing children very young." Hence their "education is greatly neglected" by their poverty-stricken parents who "can scarce afford to maintain them." As "soon as they are able to work," the children "must [therefore] apply to some trade by which they can earn their subsistence" by performing a few simple operations (Smith 1982c, 539-540, 539; and 1981, 784, 784-785). Thus their minds become just as torpid as those of their parents. When the members of any such cohort reach maturity and become parents, the same fate will await their offsprings and so on to following generations. Because he viewed this evil as cumulative, Smith ruled out any possibility that workers could regain the ability to make work-related inventions.

The milieu and manner in which potentially inventive philosophers are reared are different. They were born to families of "some rank and fortune" who were able to provide them with the best education available to their class. Being "properly educated," the potential practitioners of exogenous invention, who "have a good deal of leisure," are therefore in a position to "perfect themselves" for the acquisition of that special type of "useful knowledge" that makes them technologically adequate for the making of exogenous inventions that may have applications in a particular "branch" of manufacture (Smith 1981, 784).

Smith hinted at the possibility that the process of endogenous inventions might be revived through the education of working-class children. Hence he recommended that the "publick" establish "in every parish or district a little school, where children may be taught ... the most essential parts of education," namely to "read, write, and account." When this comes to pass, "part of [the improvements] of the machines made use of in ... manufactures" may again emanate from "inventions of common workmen" (Smith 1981, 785, 20).

Once a critical mass of inventions has been attained because "[m]ore heads are occupied in inventing the most proper machinery" than ever before, a "variety of new machines come to be invented." When that happens, still "more [are] likely to be invented," said Smith (Smith 1981, 104, 277, 104). Thus he was fully aware of the fact that "the more tools there are, the greater is the number of potential combinations" (Ayres 1944, 119). But Smith was also cognizant of the fact that in and by itself, an invention has no economic effect.

SMITH ON INNOVATION

An invention will be economically significant only if it is innovated in the sense of being applied and used in a process of production. Smith was convinced, however, that in order to flourish, innovations must be made in an institutional setting in which "the obvious and simple system of natural liberty" has been successful in keeping at bay any alternative institutional arrangement in which the government has been given "the duty of superintending the industry of private people." By natural liberty, Smith understood a system of modified laissez faire. In the first place, the government has "the duty of erecting and maintaining certain [socially beneficial] publick works and certain publick institutions" that private enterprise will not touch because of their inability to return a "profit." These included "public works which facilitate ... commerce, such as good roads, bridges, navigable canals, harbours, &c." as well as "institutions for the education of the youth" (Smith 1981, 687, 687-688, 687, 724, 758). These pronouncements led Milton Friedman to accuse Smith of providing "arguments for the interventionist and statist" (Friedman 1976, 10). Second, the government has "the duty of establishing an exact administration of justice." That is so because justice "is the main pillar that upholds the whole edifice." Hence Smith identified the system of natural liberty thus: "Every man, *as long as he does not violate the laws of justice*, is left perfectly free to pursue his own interest his own way, and to bring both his industry and capital into competition with those of any other man, or order of men" (Smith 1981, 687; 1982a, 86; and 1981, 687; emphasis added). It has therefore been argued that "the system of political economy which Smith designates 'natural liberty' is virtually synonymous with justice" (Young 1997, 125).

In the unregulated markets of such a just society, profit-seeking undertakers are forced by the "invisible hand" of competition to engage in a ceaseless process of innovation in order to survive economically. In so doing, each undertaker is lead "to promote [unintentionally] the publick interest" by "render[ing] the annual revenue of the society as great as he can" (Smith 1981, 456). There are, however, three types of activities in which men of "business" participate *qua* undertakers in the *Wealth of Nations* (Smith 1981, 456, 79). One type emanates from *revolutionary* entrepreneurial functions, another one from *imitative* entrepreneurial functions, and a third one from *routine* entrepreneurial functions. The first function is labeled revolutionary because it is the principal, and direct, cause of "great revolution[s]" in productive processes, each of which "gives occasion" to increases in "publick opulence" (Smith 1981, 422, 25; and 1982c, 497).

Those undertakers who revolutionize the economy are "bold" and daring individuals who are eager to take "extraordinary" or "superior risk" (Smith 1981, 411, 432, 33). They run such risks when they acquire new and untried inventions from the tribe of inventing philosophers. Typically, these inventions are in the form of "new machines" that promise to serve as efficient instruments for the task of "facilitating and abridging ... [the] operations" of each of those workmen who are in the employ of the entrepreneurs in question. If effected, the abridgment of labor manifests itself in the form of an increased "division of labour." Consequently, the "quantity of materials which the same number of people can work up, increases in a great proportion as labor comes to be more and more subdivided; and as the operations of each workman are gradually reduced to a greater degree of simplicity" (Smith 1981, 277). When this happens in the shop of a particular pace-setting undertaker, his marketable output increases and his unit cost of production declines. As Smith saw it, such an entrepreneur is convinced, however, that his enlarged supply will not increase perceptibly the total "quantity brought to [the relevant] market." Hence given the current total "effectual demand" for the article in question, the undertaker concludes that the "market price" for his enlarged output will not "sink" (Smith 1981, 74). With growing sales and a greater spread between unit cost and price, the bold undertaker's profits increase substantially.

How does a bold undertaker obtain the means for the innovation of a particular new invention? By accumulating a "capital" out of that "surplus part of the produce of his own" enterprise that "is over and above his own [and his family's] consumption" (Smith 1981, 279, 31). In other words, the surplus is the result of the individual undertaker's "parsimony." Smith emphasized, however, that surpluses originate in profit and that profit is a "deduction from the produce of ... labour" (Smith 1981, 337, 83). This is what Smith said on the subject:

> In all arts and manufactures the greater part of the workmen stand in need of a master to advance them the materials of their work, and their wages and maintenance till it be completed. He shares in the produce of their labour, or in the value which it adds to the materials upon which it is bestowed; and in this share consists his profit (Smith 1981, 83).

Smith considered this extraction of labor-created profit by the undertakers to be fundamental in the entrepreneur economy because it made possible those entrepreneurial savings the making of which he viewed as the first step in the process of innovation. The following is Smith's eulogy to saving:

> Parsimony, and not industry [i. e. dexterity, skills, and industriousness on the part of the agents of production], is the immediate cause of the increase of capital. Industry, indeed, provides the subject which parsimony accumulates. But whatever industry might acquire, if parsimony did not save and store up, the capital would never be the greater (Smith 1981, 337).

And, Smith might have added, innovations would be stymied so that no "additional quantity of industry" would be put "into motion" and hence no "additional value [would be give] to the annual produce" of the nation (Smith 1981, 337).

But if a system of natural liberty were established, the invisible hand of competition would force the profit-seeking bold entrepreneurs to be parsimonious so that they might have available the wherewithal for innovation when the same unseen hand impels them to be on the prowl for newly invented machines that would "enable one man [on their work force] to do the work of many." In other words, the individual pace-setting undertaker "endeavours" to furnish his workmen "with the best machines which he can ... afford to purchase." This will enable him to make "the most proper distribution of employment" among his workers so that they will "produce as great a quantity of work as possible" (Smith 1981, 17, 277).

Thus the process of revolutionary innovation begins with the act of saving on the part of bold entrepreneurs and ends when they occasion the initial economic application of the "useful machines" that their savings enable them to acquire (Smith 1981, 279). It seems, therefore, that Smith conceived of the act of revolutionary innovation in terms of four operations, or processes: capital accumulation, acquisition of an invention through the expenditure of the accumulated capital, installation of the acquired invention and, finally, start-up of the equipment in question.

The bold undertakers do not engage in these activities in a haphazard fashion. On the contrary, they make "*plans* and [design] projects" in order to "regulate and direct all the most important operations of labour, and profit is the end proposed by all those plans and projects." Because of this careful planning of the employment and distribution of labor in the process of innovation and beyond, the pioneering undertaker's expectation of extraordinarily high rates of profit is realized. As he enlarges the scale of his operations in a succession of innovations, "not only the number of profits increase, but every subsequent profit is greater than the foregoing" (Smith 1981, 266, 68; emphasis added).

In Smith's world of perfect liberty, and hence "competition," the relatively small number of bold undertakers in a given trade cannot "long enjoy their extraordinary profits without any new rivals. Secrets of this kind... can seldom be long kept" (Smith 1981, 74, 77). Once the secrets are out, hordes of imitating undertakers copy the innovations of the bold entrepreneurs. Although the imitators, who are as "timid ... [as] a country gentleman" (Smith 1981, 411), do not dare to take the risk inherent in innovation, the profits garnered by the pioneers prompt them to engage in what appears to be risk-free replication of the bold undertakers' innovations. Consequently, the production of the article in question increases with the result that "the quantity brought to market exceeds the effectual demand." The "market price will [therefore] sink" (Smith 1981, 74). At the same time, the demand for labor is increased by the spurt of activity engendered by the imitating undertakers. Hence wages go up. "The same cause, however, which raises the wages of labour, the increase in stock, tends to increase the productive power of labour." That is so, of course, because a large proportion of accumulated capitals is poured into innovation and imitation. A "smaller quantity of labour [can now] produce a greater quantity of work." There are "many commodities, therefore, which, in consequence of these improvements, come to be produced by so much less labour than before, that the increase in its price is more than compensated by the diminution of its quantity." Thus although "extraordinary profits" may be wiped out, the "ordinary rate" is preserved by the increase in productivity so that the undertakers in question are not tempted to "withdraw a part of their ... stock" from their firms (Smith 1981, 104, 77, 46, 75).

In this situation, the bold undertakers engage in a new round of innovations in the hope that they will once more be able to reap extraordinary profits. They are again successful initially, but the profit rate is soon forced down to the acceptable minimum by the imitative activities of the timid undertakers. In the long run, market "prices of all commodities are [therefore] continually gravitating" toward the "natural price" which is the minimum price that "must be paid in order" for the undertakers to receive the ordinary rate of profit (Smith 1981, 75, 73). It means that the workers are the gainers. As Smith put it, the "liberal reward of labour," and hence the "greatest public prosperity," is always and everywhere the "effect of increasing wealth" (Smith 1981, 99), an increase that is caused by the timid undertakers' imitations of the bold entrepreneurs' innovations of those exogenous inventions that were made previously by philosophers.

In addition to innovating and imitating, undertakers perform certain routine entrepreneurial activities that Smith identified as "inspection and direction" of their enterprises. He observed, however, that in "many great works, almost the whole labour of this kind is committed to some principal clerk" (Smith 1981, 66). Thus the bold and timid undertakers can concentrate their attention on the tasks of innovating and imitating, respectively. It seems, therefore, that Smith did more in his *magnum opus* than merely making "[i]ndirect references to the entrepreneurial role" (Hebert and Link 1988, 47).

CONCLUSION

Could contemporary students of the economics of innovation and entrepreneurship benefit from a visit to the *Wealth of Nations*? My answer is "yes."

In the first place, Smith demonstrated that a scholar's value judgments color his view of the world, that this view conditions the questions he asks, and that these questions shape the kind of analysis he undertakes. Consequently, Smith's message is clear: State your value judgments up front.

Second, Smith made a strong case for the proposition that "[i]nvention and innovation are both important technological and *economic* activities" (Parayil 1991, 87; emphasis added). Hence he placed inventions within the economic system and thereby within his analysis of that system. Joseph A. Schumpeter, on the other hand, maintained that "invention does not necessarily induce innovation." Consequently, invention "produces ... no economically relevant effect at all." Stressing "the element of invention" would therefore be "stressing an element without importance to economic analysis" (Schumpeter 1964, 59). Smith's approach can, therefore, serve as a corrective to Schumpeter's "unwarranted ... banishing of invention from the realm of economic activities" and from the economic analysis appertaining thereto (Parayil 1991, 87).

Third, as implied in the above discussion, Smith did not make his inquiries in order to satisfy idle curiosity. On the contrary, he did so in order to contribute to an institutional realignment of society. Thus he thundered: ""break down the exclusive privileges of corporations," i. e. trade associations, and rescind "all those laws which restrain ... competition." Once this feat has been achieved, a just "system of natural liberty" will emerge in which inventors and innovators will be stimulated to interact in such a fashion that the greatest possible welfare will be bestowed on the greatest number of people (Smith 1981, 470, 79, 687).

The issue is not the concrete policy recommendations that Smith advanced. Rather, the issue is that by formulating his suggestions for policy, Smith took the consequences of his value judgments and train of analysis. By playing the role of political economist, he demonstrated that entrepreneurial economics must be a policy science. In so doing, he filled another lacuna in Schumpeter's work on innovations, namely a "lack of policy prescriptions" (Parayil 1991, 79n).

Of course, neither Smith nor Schumpeter provided the final word in the economics of entrepreneurship and innovation. Nevertheless, those economists who work within a framework that has a Schumpeterian hue could strengthen their analyses by paying heed to Smith's pronouncements on the topics of invention, entrepreneurship and innovation.

ENDNOTES

1 All references to Smith's works are to the Glasgow Editions as reproduced photographically from the Oxford Press editions of 1976, 1978 and 1987 by the Liberty Press/Liberty Classics. The two volumes of the *Wealth of Nations* are paginated consecutively.

2 As explained below, Smith included the inventors among "those who are called philosophers or men of speculation" (Smith 1981, 21).

3 By Post-Ayresian institutionalism, I mean a type of institutionalism that goes beyond the institutionalism crafted by C. E. Ayres (1891-1972) through a synthesis of the works of Thorstein B. Veblen and John Dewey.

4 For example, Smith referred to the "necessary machinery" in his explanation of the processes in a pin factory. Moreover, in his discussion of value added at each step in the production process from the raising of sheep to the production of a "woollen coat," he emphasized the construction and use of "such complicated machines as... the mill of the fuller, or even the loom of the weaver" as well as "that very simple machine, the shears with which the shepherd clips the wool" (Smith 1981, 15, 22, 23).

5 According to John Rae, Smith's nineteenth-century biographer, businessmen in the Glasgow area founded "iron works," a "tannery," a "linen print field," copper and tin works, and a "pottery." All this occurred in the period 1742-1763 when Smith was associated with the University of Glasgow, first as a student and later as a faculty member. Hence he was "not only a teacher in Glasgow, he was also a learner, and the conditions of the time and place were most favorable, in many important ways, for his instruction" (Rae 1965, 89, 97). Recently, James A. Gherity concluded that existing evidence "is quite consistent with the [Raeian] hypothesis that Smith's views were, in fact, influenced by the economic environment" in the Glasgow area (Gherity 1992, 3671). Smith was also "fully aware of the metal-working activities of the Midlands" and of the "large-scale iron works at Carron" (Hollander 1973, 105). Thus he observed that because of "improvements of ... manufactures," there had "lately been a considerable increase in the demand for labour ... [in] Glasgow, Carron, Ayrshire, &c. " (Smith 1981, 94). It seems, therefore, that Smith's own observations confirm the interpretations of Rae *et al.* It should be pointed out, however, that Charles P. Kindleberger (1976) and Asa Briggs (1976) disagreed with Gherity, Hollander, and Rae.

REFERENCES

Ayres, C. E. *The Theory of Economic Progress*. Chapel Hill: The University of North Carolina Press, 1944.

______. *The Industrial Economy*. Boston: Houghton Mifflin Co., 1952

Briggs, Asa. "Comment." In Thomas Wilson and Andrew S. Skinner, eds., *The Market and the State: Essays in Honour of Adam Smith*. Oxford: Oxford University Press, 1976, pp. 25-33.

Friedman, Milton. *Adam Smith's Relevance for 1976*. Ottowa, IL: Green Hill Publishers for the International Institute for Economic Research, 1976.

Gherity, James A. "Adam Smith and the Glasgow Merchants." *History of Political Economy* 24 (Summer 1992): 357-368.

Hartwell, R. M. "Comment." In Thomas Wilson and Andrew S. Skinner, eds., *The Market and the State: Essays in Honour of Adam Smith*. Oxford: Oxford University Press, 1976, pp. 33-41.

Hebert, Robert F. and Albert N. Link. *The Entrepreneur: Mainstream Views & Radical Critiques*. 2nd ed. New York: Praeger, 1988.

Hollander, Samuel. *The Economics of Adam Smith*. Toronto: University of Toronto Press, 1973.

Kindleberger, C. P. "The Historical Background: Adam Smith and the Industrial Revolution." In Thomas Wilson and Andrew S. Skinner, eds., *The Market and the State: Essays in Honour of Adam Smith*. Oxford: Oxford University Press, 1976, pp. 1-25.

Mossner, Ernest Campbell and Ian Simpson Ross. "Footnote to Adam Smith's Letter to William Strahan of October 26, 1780." In Ernest Campbell Mossner and Ian Simpson Ross, eds., *Correspondence of Adam Smith*. The Glasgow Edition. Indianapolis: Liberty Press/Liberty Classics, 1987, pp. 248-249.

Parayil, Govindan. "Schumpeter on Invention, Innovation and Technological Change." *Journal of the History of Economic Thought* 13 (Spring 1991): 78-89.

Pesciarelli, Enzo. "Smith, Bentham, and the Development of Contrasting Ideas on Entrepreneurship." *History of Political Economy* 21 (Fall 1989): 521-536.

Rae, John. *Life of Adam Smith*. New York: Augustus M. Kelley 1965 (1895).

Schumpeter, Joseph A. "Science and Ideology." *The American Economic Review* 19 (March 1949): 345-359.

______. *History of Economic Analysis*. New York: Oxford University Press, 1954.

______. *Business Cycles*. Abridged by Rendigs Fels. New York: McGraw-Hill, 1964 (1939).

Smith, Adam. "Early Draft of Parts of The *Wealth of Nations*," 1763. In R. L. Meek, D. D. Raphael, and P. G. Stein, eds. *Lectures on Jurisprudence by Adam Smith*. Glasgow Edition. Indianapolis: Liberty Press/Liberty Classics, 1982, pp. 562-581.

______. *An Inquiry into the Nature and Causes of the Wealth of Nations*, 2 vols. Glasgow Edition, eds. R. H. Campbell and A. S. Skinner. Indianapolis: Liberty Press/Liberty Classics, 1981 (1776).

______. *The Theory of Moral Sentiments*. Glasgow Edition, eds. D. D. Raphael and A. L. Macfie. Indianapolis: Liberty Press/Liberty Classics, 1982a (1759).

______. *Essays on Philosophical Subjects*. Glasgow Edition, eds. W. P. D. Wightman and J. C. Bryce. Indianapolis: Liberty Press/Liberty Classics, 1982b (1795).

______. *Lectures on Jurisprudence*. Glasgow Edition, eds. R. L. Meek, D. D. Raphael, and P. G. Stein. Indianapolis: Liberty Press/Liberty Classics, 1982c (1762-1763, and 1766).

Stewart, Dugald. "Account of the Life and Writings of Adam Smith, LL.D.," 1794. In *Essays on Philosophical Subjects*, by Adam Smith. Glasgow Edition, eds. W. P. D. Wightman and J. C. Bryce. Indianapolis: Liberty Press/Liberty Classics, 1982 (1795), pp. 269-351.

Veblen, Thorstein. *The Higher Learning in America*. New York: Sagamore Press, 1957 (1918).

Young, Jeffrey T. *Economics as a Moral Science: The Political Economy of Adam Smith*. Cheltenham: Edward Elgar, 1997.

7

SCIENCE, TECHNOLOGY AND INNOVATION: ISSUES FOR THE NEOSTRUCTURALIST APPROACH TO THE SOCIOECONOMIC DEVELOPMENT IN LATIN AMERICA

Dilmus D. James
University of Texas at El Paso

That the accumulation of technological capacity is a prerequisite for sustained socioeconomic development, I will take as a given, although the notion is insufficiently appreciated in some quarters.[1] There are numerous indications that technology capacity-building in Latin America remains a marginal consideration. In this paper I will claim that, if the role of science, technology and innovation is to move closer to center stage in Latin America, neostructuralists will need to play a major role. Neostructuralists are building on structuralist foundations, while accepting some significant insights from neoclassical economics, a situation which does not bode well – neither school of thought gets acceptable marks for their understanding of the process of technology capacity-building. Yet, I will attempt to demonstrate that several important elements and interests associated with current neostructuralist thought offer convenient bridging mechanisms between their agenda and the fostering of viable and useful scientific, technological and innovative activities. The paper continues in Section I by asserting that, although there are bright spots here and there, the contributions of science and technology to socioeconomic development in Latin America are suboptimal. In Section II the neostructuralists are identified and a case is made that, by default if for no other reason, their contributions are direfully needed for fostering the accumulation of internal technology capacity.

Section III discusses several areas of science and technology policy that need to be addressed, most of which should be compatible with current neostructuralist strengths and concerns. A short section concludes the paper with some reflections as to what actions might further the cause of explicitly incorporating science, technology and innovation policy into the broader panorama for fostering socioeconomic progress.

SCIENCE, TECHNOLOGY AND INNOVATION: BRIGHT SPOTS; OVERALL WEAKNESS

Fortunately, it is not difficult to identify bright spots on the scientific, technological and innovational fronts in Latin America.[2] The Technological Innovation Center (Centro para la Innovacion Tecnologica, or CIT) at the National Autonomous University of Mexico (Universidad Nacional Autonoma de Mexico, or UNAM) is often praised for effectively linking R&D with the productive sector. A small institute in Chile, The Corporation for the Development of Capital Goods (Corporacion para el Desarrollo de Bienes Capitales), is credited with contributing a significant technological upgrading of the capital goods industry in that country. The early experience of several Brazilian technology parks combining agglomerations of related high-tech enterprises with universities and research institutes appears is producing dozens of products including a vaccine to combat hepatitis B.[3] The National Institute for Agricultural Technology (Instituto Nacional de Tecnologia Agricultura) in Argentina, through a biotechnology technique called "balanced lethal fretons," has succeeded in developing new commercially viable strains of maize (Katz and Bercovich 1988).[4] Both the Electrical Research Institute of Mexico (Instituto de Investigacion Electrica de Mexico) and the Mexican Petroleum Institute (Instituto Mexicano del Petroleo) on which it was modeled have good track records for in-house innovation and promotion of innovation externally.[5] Sophisticated informatics systems for the financial sector have been developed in Brazil where it flourished domestically and in the export market. The quasi-public cooperative, CONAPROLE, in Uruguay, partly by increasing access to technology for small and medium-scale producers in the dairy industry, and forging and strengthening linkages among them, helped make Uruguay the largest dairy exporter in Latin America. The catalog could easily be extended – for a sampling of cutting-edge scientific projects in Latin America, see the special section on the topic in *Science* (AAAS 1995). Case studies are available on enterprises with exemplary records of technology capacity-building.[6]

Despite bright spots, all is not well. In a recent review of science and technology policy and science and technology indicators in Latin America, Carlos Correa (1995) concluded that overall science and technology budgets are modest and declining, there is a low patent activity, and a shortage of scientific and engineering personnel engaged in R&D. Like virtually every other observer, he remarks on the relative overemphasis on research, as opposed to development and to the lack of linkages between R&D and the productive sector. Science and technology policy is not properly coordinated with industrial, fiscal, financial, trade and other policies. More disturbing evidence comes from Paolo Guerrieri who examined trends in the exports of Argentina, Brazil and Mexico, countries that account for 80 percent of Latin American trade. While primary resource-based products and scale-intensive industries (steel products, basic chemicals and transportation equipment have been doing very well in international markets, both specialized supplier goods and science-based products, after a substantial growth during the 1970s and early 1980s, have, as a proportion of world trade, lost ground rapidly. Unfortunately, these last two categories "are the major sources of technical change and technological inputs to the productive system as a whole" (Guerrieri 1994, 178).

National studies of Argentina, Brazil and Mexico are also disquieting. Technological change in Argentina's manufacturing sector is largely confined to minor innovations, and moreover, there is virtually a stigma associated with academic scientists who become involved with applied research (Katz and Bercovich 1993). It is difficult to believe that the accumulation of national technological capacity is close to the center of Brazil's policy concerns in view of a tripling of undergraduate university enrollment during the 1970s while full-time faculty actually declined (Dahlman and Frischtak 1993). An Organization for Economic Cooperation and Development (OECD) (1994) study on Mexico concluded that science and technology policy instruments are outdated and fragmented, considerations suggesting "that profound structural change is required for the Mexican science and technology system" (OECD 1994, 53). Wionczek and Marquez (1993) are, perhaps, even more pessimistic. After reviewing recent science and technology policy and the environment in which it operates, they conclude "All these circumstances seem to reinforce domestic and foreign doubts about the feasibility of a reasonably coherent and long term science and technology policy in Mexico" (49).

THE NEOSTRUCTURALISTS

There are a number of views and approaches that distinguishes neostructuralists form structuralists and neoliberals. Indeed, from Bitar (1988), Ffrench-Davis (1988), Lustig (1991), Rosales (1988), Sunkel (1993), Sunkel and Ramos (1993), and Sunkel and Zuleta (1990), one could compile an extensive list, but the following six distinguishing characteristics will suffice for our purpose.

First, neostructuralists continue the structuralist tradition of stressing structural shortcomings, many of which constitute supply bottlenecks or asymmetries in power and social status. More than market forces or marginal adjustments are required to

> (1) promote or stimulate missing markets (long term capital markets and futures markets in foreign exchange); (2) strengthen incomplete markets (the market for technology); (3) eliminate or correct structural distortions (the heterogeneity of the production structure, the concentration of property, the segmentation of the capital and labor markets); and eradicate or compensate the most significant market imperfections arising from economies of scale, externalities, and learning (arising from technology and trade), among others (Sunkel and Ramos 1993, 7).[7]

Second, and also compatible with the structuralist tradition, political, social and cultural conditions, rather than being "given" or exogenous, are treated as an integral part of economic analysis and policy formulation. This is one reason why historical expositions are so common in neostructuralist writing. The "holistic" approach is well illustrated by neostructuralist treatment of "inertial inflation," whereby expectations by, and conflicts among, economic agents, a product of past experience, means that "in addition to monetary and fiscal policies and structural changes, measures should be jointly formulated on the psychological, social and politico-institutional aspects" (Bitar 1988, 48).

Third, the state has a crucial role to play. This does not necessarily imply a big state, but it implies an active and strong, "lean and mean" configuration. The government sector, however, must do better at setting priorities and avoiding overextending its reach. The state and private sectors should be complementary rather than adversarial. Solid neostructuralist-oriented discussions of the role of the market vis-a-vis the state can be found in Gerchunoff and Torre (1992) and Salazar-Xirinachs (1993).

Fourth, although not true of nascent neostructuralism in the 1970s, later and current neostructuralists are keenly interested in forging meaningful links between short- and medium-term policy concerns with long-run considerations. Perhaps contemporary neostructuralists are giving the long pull more attention than even the structuralists, and certainly the focus on the long term surpasses that of the neoliberal camp.

Fifth, neostructuralists share the neoliberal position that internal and external economic stability are important goals and appropriate monetary and fiscal instruments need to be brought to bear to achieve a degree of stability of the domestic price level and balance of payments. Furthermore, neostructuralists agree that macroeconomic stabilization and adjustment policies were necessary after the debt crisis in the early 1980s, yet they insist that the "let the chips fall where they may" attitude of the neoliberals put unnecessary burdens on domestic production and capital formation, as well as furthering disparities in incomes between the haves and have nots.[8] Among other measures, they favor selective interventions to facilitate the transfer of resources from the non-tradable sector to tradables, to place the burden of the decline in imports on consumption items rather that on intermediate inputs and capital goods, and to provide levels of government revenue that can maintain expenditures on necessary infrastructure projects.

Sixth, the older inward-looking economic model exhausted itself as domestic production replaced manufactured imports beyond any reasonable point, and global liberalization threatened to relegate inward-looking economies to a state of economic and technological stagnation. Neostructuralists recognize that insertion into the international competitive market is necessary to change entrepreneurial behavior, realize economies of scale, increase exposure to technical learning, among other benefits. There is, however, no intrinsic conflict between a simultaneous pursuit of import substitution and export-promoting measures.[9]

With those cited in this section, all neostructuralists, we have a sizable roster; Ramirez (1993) includes others in his list, more, not yet cited, are included in the

Sunkel edited collection (1993), and we can safely include many other economists associated with the Economic Commission for Latin America and the Caribbean (ECLAC). While by no means monolithic on any single issue, neostructuralists constitute a fairly cohesive group of social scientists. To borrow a British expression, I will now attempt to lay another plate on their back.

James Street and I (1982) once alleged structuralists did not spend enough effort on the role of science and technology in promoting socioeconomic development, and although Osvaldo Sunkel pointed out that structuralism always stressed industrialization, "which is seen as the bearer of technology," he agreed that technology "is an aspect of development theory that has been badly neglected by structuralism" (Sunkel 1989, 526). It will be lamentable indeed if neostructuralism follows a similar path because, as I will contend, there is no other group better situated to pull science and technology into the mainstream of policy in Latin America. As we will see shortly, Latin America has suffered no dearth of first-class studies on science and technology policy, and these investigations did occasionally influence the course of events, but in general, efforts have been too fragmented and isolated – there developed no cohesive school of economic thought giving suitable weight to science and technology policy in Latin America.

Why do I not implore the neoliberals to do the job? Because it is hopeless, that is why. After several wake-up calls such as the realization that technology loomed large in accounting for the "residual" that kept cropping up in studies on economic growth during the 1950s, and the current frenzy over achieving and maintaining international competitiveness, orthodox economists are devoting increasing effort to technology, but their methodology constitutes an enormous handicap. While true that mainstream economists have been responsible for devising the "new growth" or "endogenous growth" theory, which recognizes the constructive influence of investment on technological progress, the analysis remains highly abstract and reductionist.[10]

As Wassily Leontief writes of econometric models of traditional economics: "In these models, increasingly ingenious but still utterly unreliable methods of indirect statistical inference are employed in vain to compensate for the lack of hard, systematically organized factual information."[11] Elsewhere he remarks on the "continuous unwillingness of academic economists to give up their traditional reliance on abstract mathematical formulas – linked to reality only by a very fragile bridge of indirect statistical inference ..."[12] Pretty damning words from a mathematical economist. Aside from a debilitating association with neoclassical thought, neoliberals are singularly closed to heterodox advise.[13]

There are two schools of thought that do put science, technology and innovation very close to the epicenter of economic development, namely the institutionalist and neoSchumpeterian schools. In my opinion the neoSchumpeterians are doing the most exciting and original work in the field,[14] but the brunt of their effort is focused on developed countries, and when developing countries are investigated, it is almost exclusively geared to top-tier nations of the Third World. They may heavily influence Latin American efforts, but they are not likely to lead the charge in the region.[15] As to institutionalists, although it pained me to say so, elsewhere I have ventured the opinion that "they have not been sufficiently influential in the field of development economics beyond their own circle. Although there are notable exceptions like Wendell Gordon, James Street and Thomas DeGregory, there has never been a significant core of institutionalists who concentrated their energies on the role of science and technology in developing countries" (James 1996, 143). In addition the Thorstein Veblen-Clarence Ayres axis of institutionalism, the most important foundation for stressing technology, while inspirational and insightful, failed to lead to a sizable body of practical studies specifically aimed at developing and advocating policy packages designed to foster science, technology and innovation for development purposes. Institutionalism, of course, is not devoid of work of this sort, I could cite, illustratively, the work of Paul Strassmann on the housing and the construction industries that often has a technological dimension;[16] work by William E. Cole and associates at the University of Tennessee at Knoxville on the continuous improvement firm; and with collaborators, I have been trying to make a modest contribution with a series of studies, at the University of Texas at El Paso, that are related to Mexico.[17] Still, though institutionalists also have a considerable contribution to make, there is little possibility that they can effect a major cataclysm in the contours of science and technology policy in Latin America.

What I am asserting is that if any group is to elevate considerations of scientific, technological and innovational pursuits to their proper position in Latin American development, neostructuralism is the only kid on the block. If they are not willing to embrace the challenge, no other group is waiting in the wings.

WHY NEOSTRUCTURALISTS?

Fortunately, neostructuralists are imbued with many assets that render them up to the job. The prestige and influence of a well recognized and well regarded group that is willing to expend considerable energy, persistently and effectively, is required to move technology capacity-building very near to the heart of development policy. A great deal of first rate work has been done on science and technology and Latin American development, a sampling of which will be mentioned below, yet, evidently, this corpus of literature has not sufficiently influenced the center of gravity of development policy. The neostructuralists, however, have the requisite clout.[18] Furthermore, the regard that neostructuralists have for long-run goals and the influence of historical forces is compatible with the lengthy, cumulative, evolutionary process required for the accumulation of technological capacities. Well before the onslaught of neoliberalism, structuralists understood that to have any chance for achieving sustained socioeconomic development depends on devising policies that are, essentially "home grown." This is the significance of "Development from Within" in the title of the Sunkel edited volume (1993). Scientific and technological progress and the associated socioeconomic consequences take place within different cultural, social, political, and economic contexts – a universalist approach of "getting prices right" has serious, if not fatal, limitations. Like institutionalists, neostructuralists are acutely aware of this. A somewhat related point: fostering technology capacity-building in Latin America requires state intervention in such a way that will maximize private-sector participation, something that neostructuralists are learning to accomplish. Finally, neostructuralists are fully cognizant of the costs that attend upheaval and disruption that stem from economic and social inequities. Some more specific bridging mechanisms follow.

Literature on Science, Technology and Innovation

The literature on science, technology and innovation constitutes a solid platform for further inquires and analyses. One might mention the series of studies, Estudios sobre el Desarrollo y Tecnologico, by the Regional Program on Scientific and Technological Development of the Organization of American States during the 1970s that inventoried science and technology resources in many Latin American countries; the thorough examination of science and technology policy in Latin America that was part of the Science and Technology Policy Instruments project sponsored by Canada's International Development Research Centre.[19] and the work on plant-level innovations, with funding from a consortium of United Nations agencies, by Jorge Katz and associates during the last half of the 1970s and early 1980s.[20] I have already alluded to several recent studies of national systems of innovation.[21] Work on imperfections in international technology markets by Constantine Vaitsos palpably influenced science and technology policy;[22] the role of technical consultants in Latin American development by Araoz and associates (1981) remain, in my opinion, unsurpassed; and Mexico's *Plan Indicativo*, though ill fated, was highly regarded in the Third World.[23]

The debates on the potential role of Latin American universities in national innovation systems have provided many insights. The literature ranges from Jorge Sabato's exhortations to include the scientific-technological community as one vertex of a "triangle" with the state and the productive sector (Sabato 1969),[24] through a period of discouragement when many gave up on universities, to guarded optimism.[25] Simon Schwartzman and Hebe Vessuri, who were prominent in these debates, have also made major contributions to the understanding of how political and sociological conditions impinge on the pace and direction of scientific and technological activity.[26] Over the years the Inter-American Development Bank has sponsored some useful investigations.[27]

There is a voluminous literature on the informatics industry in Latin America, especially concerning Brazil.[28] The technological capacity of many other industries has been reviewed; the Mexican studies on textiles and garment production by Alfonso Mercado (1980) and Gerard K. Boon (1984), the chemical industry by Jose Giral B. et al. (1978), and the petrochemical industry by Soto Angli and Castillo (1983) come to mind. Pioneering works like those of Sercovich (1975) and Wionczek (1971) explored the technological repercussions of foreign investment. The past decade has seen two review essays on science and technology policy in the *Latin American Research Review*.[29] A welcome recent collection of essays on the political economy of science in Latin American (Bastos and Cooper 1995) has appeared under the sponsorship of the United Nations Institute for New Technologies. Several Latin Americans, aside from those already indicated, have long and distinguished lists of accomplishments in the area of science and technology policy and/or the socioeconomic change of technical change, including Daniel Chudnovsky, Amilcar Herrera, Francisco Sagasti, Francisco Sercovich, and Simon Teitel.

The list could easily be extended, but this selective sampling of literature confirms the existence of a sound tradition of research on science, technology and innovation in Latin America – the problem is that there is too little interaction between this research community and the neostructuralists. The latter group has been subjected to "crowding out" whereby analysis of macroeconomic crises has tended to usurp most of their energies. This is exacerbated by career "lock in" entailing considerable personal and professional cost in altering the trajectory of one's professional specialization. However, there are interrelations and overlaps – the group investigating science and technology and neostructuralists are not like two ships passing in the night, rather the situation more closely resembles passing during the first rays of dawn. We can be fairly confident that neostructuralists are familiar with much of the literature alluded to earlier. As an example, Osvaldo Sunkel, certainly a leading neostructuralist,[30] wrote early in his career a frequently cited article on technology and the Latin American university (1971) and has retained an interest in technology and development, although his research has gone in other directions. Much the same can be said of Victor Tokman (1974, 1981). Technology gaps are frequently mentioned by neostructuralists as one of the structural bottlenecks – Ramos and Sunkel, for example, refer to incomplete markets for technology and "the most significant market imperfections arising from economies of scale, externalities, and learning (arising form technology or trade) ..." (1993, 7).

There are three more recent examples of neostructuralist-oriented writing, all very impressive, in my opinion. Ennio Rodriguez (1993) has a chapter in the Sunkel collection wherein he makes a persuasive case that achieving endogenous mastery over technology should be a major goal of Latin American development policy. In elaborating his argument Rodriguez relies mostly on literature from outside Latin America; one hopes that this is a prelude for follow-up investigation delving more fully into actual conditions in the region. One would not guess it, perhaps, from the title, but ECLAC (1990) produced a study which masterfully interrelates technological considerations with matters relating to social equity. The late Fernando Fajnzylber (1990) built a potent technological dimension into his comparison of Latin American economic performance with other industrialized or industrializing regions. It would be extremely beneficial if these lines of exploration could be extended and intensified.

Macroeconomics and Technology Capacity-Building

Another bridging mechanism is macroeconomics itself. There is, of course, a clear and direct relation between macroeconomic conditions and scientific, technological and innovational activities.[31] The compression of investment and imports during periods of stabilization is well documented and a point of contention with neostructuralists. To the extent that domestic investment involves embodied technical change or induces complementary disembodied technical adjustment learning, and to the extent that import flows incorporate science or technology inputs (scientific instruments, capital goods, consulting services), the cause of technology capacity-building is blunted. Cuts in R&D expenditure during the financial/economic crisis of the early and 1980s were universal among Latin American countries. Illustratively, federal government expenditure on science and technology by Mexico plummeted from US$1.45 billion in 1981 to $472 million in 1983 – it took a decade to return to the 1981 level; Venezuela saw national R&D spending go from US$278.3 million in 1983 to approximately $187 million the following year and, through 1991, the 1983 level had not been attained; and Brazil's spending on science and technology dropped from US$2.62 billion in 1982 to $2.12 billion the following year.[32] Much of the sunk costs of research projects whose funds are curtailed while in progress will never be recovered, some projects will fall below the critical minimum threshold for resources necessary to have much chance of success, youngsters are discouraged from choosing a career in science and technology, some key personnel will leave for occupations outside the science and technology sector, morale within R&D and technical institutes is adversely affected, and the uncertainty attending "stop and go" economic performance discourages innovation.

Obviously, any reduction in wide swings in macroeconomic variables would, in itself, contribute to technology capacity-building, and the neostructuralists are, to a large extent, preoccupied with devising policy packages that can ameliorate these fluctuations. Two other areas warrant attention. While the costs of reduced investment in terms of future economic growth are fairly well known, the deleterious impacts on science, technology and innovation, some of which we have just identified, have not been sufficiently documented – an effort in this direction might be useful. Second, are there reasonable measures that can be implemented that could, at least partially, shield the science and technology sector from macroeconomic fluctuations? In this regard, their familiarity with selective, as opposed to neutral, policy instruments should stand the neostructuralists in good stead.

Human Resources

The next point need not detain us for long, although the brevity belies its importance. Like the structuralists, neostructuralists have a keen and abiding interest in the development of human resources. Not only are neostructuralists advocates of more and better education; they are also responsible for penetrating analyses and solid policy prescription in this area.[33] The connection between science, technology and innovation are rather obvious. Science and technology in the general curriculum has the direct effect of imparting useful knowledge and a heightened sensitivity and appreciation of the potential contributions of science and technology. The educational system serves as a source of skilled and specialized human resources and can participate, through research, in generating new knowledge. Upgrading the educational system is given high priority in the neostructuralist scheme of things – in view of the intimate interrelationships between education and scientific and technological progress, it would seem no great leap to accord equal status to national technology capacity-building.

Similar observations are warranted regarding structuralist and neostructuralist concern with unemployment and underemployment. A major seat of this concern is the International Labor Office offshoot, Programa Regional del Empleo para America Latina y el Caribe (PREALC), which has recently been directed by a neostructuralist, Victor Tokman.[34] Technology, of course, impinges *inter alia* on labor productivity, skill profiles of the labor force, aggregate and occupational demand for employment, and training and retraining needs. PREALC has certainly given these matters attention, but with the dissolution of the Employment and Technology Branch of the parent organization in Geneva, perhaps there is more opportunity and imperative to concentrate more heavily on technology-employment interactions.

State Interventions

As mentioned earlier, fostering technology capacity-building requires state intervention to fill various voids and state guidance and encouragement to elicit the utmost from the private sector. Aside from playing a significant role in upgrading the educational system, a subject already broached, state action is needed for supplying science and technology infrastructure such as public research testing and metrology institutions, productivity centers, quality control facilities, technology extension services, data banks and advisory services relating to technical and marketing matters, and industry-specific support centers. Once up and running, financial support can be provided by clients of these types of support institutions, but the initial impetus must come from the state.

Far more amorphous, but equally important, are ways in which the state can influence the institutional environment. The state can establish a non-adversarial atmosphere conducive to R&D and innovation in the private sector; attempt to forge links between R&D (including institutions of higher education) and productive activity; encourage the formation of professional associations and other measures for facilitating learning linkages; balance the rewards from research with the social benefits from technological progress through an appropriate intellectual property rights regime; and provide or encourage financial institutions that are sources of venture capital and capable of evaluating projects incorporating novel technology. Once again, difficult though the assignment may be, the neostructuralist economists are the best bet to initiate and implement the proper policy instruments. The agenda definitely includes a program of technology for the very poor which entails primarily enabling resources for access to technology rather than generation of new technologies per se.[35]

SUMMARY AND CONCLUDING REMARKS

Technology capacity-building is on the periphery of development policy in most Latin American nations; it needs to move much closer to the center of gravity. This is despite a great deal of excellent analysis of science, technology and innovation in the region. I have argued that the neostructuralists, a fairly cohesive group with considerable influence, constitute the best hope for accomplishing this end. As a whole, neostructuralists have not gotten their feet wet in the field of science and technology, but some have, and technology is recognized as one area harboring structural shortcomings that must be overcome. Technology capacity-building will require judicious state actions with policy packages tailored to specific situations taking into account local, endogenous conditions. The overall effort should complement the goal of social equity. Neostructuralists get high marks on all of these considerations.

NeoSchumpeterians have done well at blending macroeconomic concerns with an emphasis on technology capacity-building and serve as a rough, indicative model for neostructuralists. Intensified interaction among neostructuralists, institutionalists and neoSchumpeterians is in order. The same goes for interaction between neostructuralists and those active in researching science and technology. I am convinced that considerable informal exchange takes place, but I am thinking about more formal fora and associations for devising a frontal attack on technology bottlenecks and deficiencies. Symposia, conferences, and workshops deliberately targeted to policy formulation are obvious possibilities as are collaborative research endeavors. The international community and agencies such as the Economic Commission for Latin America and the Caribbean, Inter-American Development Bank, Programa Regional del Empleo para America Latina y el Caribe and the United Nations Conference on Trade and Development can play a crucial leadership role.

ENDNOTES

1 An institutionalist friend who recently co-authored a textbook on economic development tells me that he and his collaborator received all kinds of kinds of static from neoclassical reviewers who were greatly disturbed by the emphasis on the role of technology.

2 Unless otherwise cited, these examples are found, with further references, in ECLAC (1990).

3 Other Latin American successes in biotechnology can be found in Bifani (1988).

4 For more innovative activity by this institution, see Katz and Bercovich (1993).

5 On the Mexican Petroleum Institute, see Wionczek and Marquez (1993).

6 See Dahlman and Frischtak (1993) on Metal Leve; Frischtak on EMBRAER; ECLAC (1990) on Micro Sistemas, and Girvan and Marcelle (1991) and Electric Arc Limited.

7 Ramos (1994) does and excellent job of elaborating some of these and other structural imperfections.

8 See Ramirez (1993) and Ramos (1993) for discussions of the neostructuralist position on stabilization and adjustment.

9 "Import substitution and export promotion do not have to be mutually exclusive options: the real challenge to be faced is how to choose a combination of policies which will optimize the country's own productive capacity and increase their international competitiveness" (Bitar 1988).

10 Illustratively, see Lucas (1988) and Romer (1986; 1990).

11 Leontief (1982) cited by Swaney and Premus (1983).

12 Leontief (1981) cited by Swaney and Premus (1983).

13 I am reminded of the fictional artist, Gulley Jimson, in the closing paragraph in Joyce Carey's *Through the Horse's Mouth* – speaking to a nun who was worried that he would agitate some teenagers, Jimson says "That's all right – they can't hear me because of the noise of the traffic and because they aren't listening. And it wouldn't make any difference if they did. They're too young too learn, and if they weren't they wouldn't want to."

14 Among other ideas, neoSchumpeterians are concerned with non-codified, or tacit knowledge, technical learning through user-producer relationships, scientific or technological trajectories (and relatedly, path dependent technological and innovational change), national innovation systems, technology-driven long (or Kondratief) economic cycles, and international trade partially based on absolute technological advantage.

15 Two excellent sources of neoSchumpeterian thought are Dosi et al. (1988) and OECD (1992). Although the latter gives no individual attribution to authorship, it is clearly very heavily influenced by neoSchumpeterian positions.

16 See, for example, Strassmann (1973, 1979, 1989).

17 Publications of the Cole group include Cole and Mogab (1995), Bos and Cole (1994), -- Wilson, Ballance and Pogany (1995) was significantly influenced by Cole; the work at the University of Texas at El Paso includes James (1988; 1991), Brannon et al. (1994), and James and Vickers (1995).

[18] Bitar (1988, 45) refers to neostructuralism as "the mainstream thinking of the region."

[19] See Segasti (1978) for a summary report of the STPI program.

[20] Many of the studies can be found in Katz (1987).

[21] See also Villaschi (1994).

[22] Vaitsos (1974) summarizes and expands on his earlier work.

[23] See CONACYT (1976) for the plan and James (1980) for the fate of the plan.

[24] This brief piece by Sabato is one of the earliest to focus on internal technological capacities in Latin America rather than on technology transfer.

[25] See Schwartzman (1986) in which he reviews some of the debate and describes how his position gravitated from a pessimist to one more hopeful. Vessuri (1993) elaborates some of her current thoughts on the matter.

[26] See Schwartzman (1991) and Vessuri (1987; 1993).

[27] See IBD (1983) on technology exports, IBD (1988) on new technologies, and Chudnovsky et al. (1992), Kessel and Samaniego (1992) and Agacino et al. (1992) for country studies on Argentina, Mexico, and Chile, respectively.

[28] See Bastos (1992) and Erber (1995) and citations therein.

[29] See Schwartzman (1989) and James (1993).

[30] I suppose it goes without saying that some prominent structuralists evolved into neostructuralists, Osvaldo Sunkel being one of the most foremost examples.

[31] Using Mexico as a case study, I have attempted to elaborate on the interrelations between macroeconomic instability and technology capacity-building in James (1995).

[32] See Parra Moreno and Mejia Gomez (1993), Sosa de Golindano and Pena Cedillo (1993) and Galvao et al. (1993) on Mexico, Venezuela and Brazil, respectively.

[33] I am especially impressed with Ramos (1994) and ECLAC/UNESCO (1992).

[34] See his contribution to the Sunkel edited collection on neostructuralism, Tokman (1993).

[35] See James (1994) and CSTD (1995).

REFERENCES

Agacino, G., G. Rivas and E. Roman. *Aperatura y Eficiencia Productiva: La Experiencia Chilena 1975-1989*. Washington, DC: Inter-American Development Bank, 1992.

American Association for the Advancement of Science. "Science: Points of Light in Latin America." *Science* 267: 5100(10 February 1995): 807-828.

Araoz, Alberto, ed. *Consulting and Engineering Design in Developing Countries*. Ottawa: International Research Centre, 1981.

Bastos, Maria-Ines. "State Autonomy and Capacity for S&T Policy Design and Implementation in Brazil." In Bastos and Cooper, 1993, 68-108.

________ and Charles Cooper, eds. *Politics of Technology in Latin America*. London: Routledge, 1995.

Bifani, Pablo. "Biotechnology: Overview and Developments in Latin America." In IDB, 1988, pp. 195-283.

Bitar, Sergio. "Neoconservatism versus Neo-Structuralism in Latin America." *CEPAL Review* No. 34(April 1988): 45-62.

Boon, Gerard K. *El Mercado de Tecnologia: Con Referencia a Fibras Sinteticas, Telas y Ropas*. Mexico, DF: El Colegio de Mexico, 1984.

Bos, Antonio and William E. Cole. "Management Systems As Technology: Japanese, US and National Firms in the Brazilian Electronics Sector." *World Development* 22:2(February 1994): 225-236.

Brannon, Jeffrey T., Dilmus D. James and G. William Lucker. "Generating and Sustaining Backward Linkages Between Maquiladoras and Local Suppliers in Northern Mexico." *World Development* 22:12(December): 1933-1945.

Carey, Joyce. *Through the Horse's Mouth*. New York: Harper & Row, 1944.

Chudnovsky, Daniel, Ricardo Soifer, Andres Lopez, Fernando Porta, Martina Chidiak and Pablo Sierra. *Apertura, Productividad Desarrollo Technologico en las Industrias Petroquimica, Sidurgica, Automotriz y Maquinas Herramientas en la Argentina*. Washington, DC: Inter-American Development Bank, 1992.

Cole, William E. and John W. Mogab. *The Economics of Quality Management: Clashing Paradigms in a Global Market*. Cambridge, Mass.: Blackwell, 1995.

Commission on Science and Technology for Development (CSTD), Economic and Social Council, United Nations. *Science and Technology for Basic Needs: A Bridge.* Panel on Technology for Small-Scale Economic Activities to Address the Basis Needs of Low-Income Populations. Geneva: United Nations, 15 May 1995.

Consejo Nacional de Ciencia y Tecnologia. *Plan Indicativo de Ciencia y Tecnologia.* Mexico, DF: CONACYT, 1976.

Correa, Carlos M. "Innovation and Technology Transfer in Latin America: A Review of Recent Trends and Policies." *International Journal of Technology Management* 10:7/8(1995): 815-846.

Dahlman, Carl J. and Claudio R Frischtak. "National Systems Supporting Technical Advance in Industry: The Brazilian Experience." In Nelson, 1993, pp. 414-450.

Dosi, Giovanni, Christopher Freeman, Richard Nelson, Gerald Silverberg, and Luc Soete, eds. *Technical Change and Economic Theory.* London: Pinter, 1988.

Economic Commission for Latin America and the Caribbean and Regional Office for Education in Latin America and The Caribbean, UNESCO. *Education and Knowledge: Basic Pillars of Changing Production Patterns with Social Equity.* Santiago: United Nations, 1992.

Erber, Fabio Stefano. "The Political Economy of Technology Development: The Case of the Brazilian Informatics Policy." In Bastos and Cooper, 1996, pp. 196-224.

Fajnzylber, Fernando. *Unavoidable Industrial Restructuring in Latin America.* Durham, NC: Duke University Press, 1990.

Ffrench-Davis, Ricardo. "An Outline of a Neo-Structuralist Approach." *CEPAL Review* No. 34(April 1988): 37-44.

Frischtak, Claudio R. *Learning, Technical Progress and Competitiveness in the Computer Aircraft Industry: An Analysis of Embraer.* CEPR Publication No. 356. Stanford: Center for Economic Policy Research, University of Stanford, 1992.

Galvao, Antonio Carlos F., Ariel C. Garces Pares, Arthur Oscar Guimaraes, Eduardo Baumgratz Viotti and Sandra Hollanda. "Brazil." In Oro and Sebastian, pp. 63-85.

Gerchunhoff, Pablo and Juan Carlos Torre. "What Role for The State of Latin America?" In *Toward a New Strategy for Latin America: Pathways from*

Hirschman's Thought, Simon Teitel, ed. pp. 259-280. Washington, DC: Inter-American Development Bank, 1992.

Giral B., Jose, Sergio Gonzalez P. and Eduardo Montano A. *La Industria Quimica en Mexico*. Mexico, DF: Redacta, 1978.

Girvan, Norman P. and Gillian Marcelle. "Overcoming Technological Dependence: The Case of Electric Arc (Jamaica) Limited, a Small Firm in a Small Country." *World Development* 18:1(January 1991): 91-107.

Guerrini, Palao. "International Competitiveness, Trade Integration and Technological Interdependence." In *The New Paradigm of Systemic Competitiveness: Toward More Integrated Policies in Latin America*, In Colin I. Bradford, Jr., ed., pp. 171-206. Paris: Organization for Economic Cooperation and Development, 1994.

Inter-American Development Bank. "Technology Exports from Latin American." In *Economic and Social Progress in Latin America*, pp. 141-161. Washington, DC: IDB, 1993.

________. "Special Section on Science and Technology." *Economic and Social Progress in Latin America*, 1988 Report, pp. 141-161. Washington, DC: IDB, 1988.

James, Dilmus D. "Mexico's Recent Science and Technology Planning: An Outsider Economist's Critique." *Journal of Interamerican Studies and World Affairs* 22:2(May 1980): 163-193.

________. *The Impact of Technology Imports on Indigenous Technological Capacity: Case Study of Mexico*. World Employment Programme Working Paper No. 184. Geneva: International Labor Office, 1988.

________. "Capital Goods Production and Technological Learning: The Case of Mexico." *Journal of Economic Issues* 25:4(December 1991): 977-991.

________. "Technology Policy and Technological Change: A Latin American Emphasis." *Latin American Research Review* 28:1 (1983): 89-101.

________. "Basic Needs – Old Imperatives, Fresh Approaches." Paper, Second Meeting, Panel on Technology for Small-Scale Economic Activities to Address the Basic Needs of Low-Income Populations. Geneva: United Nations, 17-19 October, 1994.

________. "Economic Crises and Technology Capacity-Building: A Neglected Case of Hysteresis." Paper, Southwestern Economics Association, March 20-23 1995, Houston.

________. "Technology and Third World Development: A Twenty to Thirty Year Retrospective." *In The Institutional Economics of International Economic Relations,* John Adams and Anthony Scaperlanda, eds., pp. 143-159. Boston: Kluwer, 1996.

________ and Danny Vickers. "Computer Software in Developing Countries: A Case Study on Cd. Juarez, Mexico." *Journal of Global Information Management* 3:4(Fall 1995): 24-31.

Katz, Jorge and Nestor A. Bercovich. "Innovacion Genetica, Esfuerzos Publicos de Investigacion y Desarrollo y la Frontera Tecnologica Internacional: Nuevos Hibridos del INTA." *Desarrollo Economico* 28:110 (Julio/Septiembre 1988): 209-243.

________. "National Systems of Innovation Supporting Technical Advance in Industry: The Case of Argentina." In Nelson, 1993, pp. 451-475.

Kessel, Georgina and Ricardo Samaniego. *Apertura Comercial, Productividad y Desarrollo Tecnologico: El Caso de Mexico.* Washington, DC: Inter-American Development Bank, 1992.

Leontief, Wassily. "If Mr. Reagan's Policies Flop, Then What?" *The New York Times* 22 February 1981.

________. "The American Economy and Its Prospects." Statement submitted to the Joint Economic Committee of the United States Congress, 19 January 1982.

Lucas, Robert E. Jr. "On the Mechanics of Economic Development." *Journal of Monetary Economics* 22:1, pp. 3-42.

Lustig, Nora. "From Structuralism to Neostructuralism: The Search for a Heterodox Paradigm." In *The Latin American Development Debate,* Patricio Mellor, ed., pp. 27-42.

Mercado, Alfonso. *Estructura y Dinamismo del Mercado de Tecnologia Industrial en Mexico: Los Casos del Poliester, Los Productos Textiles y el Vestido.* Mexico, DF: El Colegio de Mexico, 1980.

Nelson, Richard R., ed. *National Innovation Systems: A Comparative Analysis.* New York: Oxford University Press, 1993.

Organization for Economic Co-operation and Development. *Technology and the Economy: The Key Relationships.* Paris: OECD 1994.

Oro, Luis A. and Jesus Sebastian, eds. *Los Sistemas de Ciencia Y Tecnologia en Iberoamerica.* Buenos Aires: Editorial Universitaria de Buenos Aires, 1993.

Parra Moreno, Javier and Daniel Mejia Gomez. "Mexico." In Oro and Sebastian, pp. 253-278.

Ramos, Joseph R. "Microeconomic Equilibria and Development." In Sunkel, 1993, pp. 83-123.

________. "Employment, Human Resources and Systemic Competitiveness" In *The New Paradigm of Systemic Competitiveness: Toward More Integrated Policies in Latin America,* Colin I. Bradford, Jr., ed., pp. 249-269. Paris: Organization for Economic Cooperation and Development, 1994.

________ and Osvaldo Sunkel. "Toward a Neostructuralist Syntheses." In Sunkel, pp. 5-19.

Rodriguez, Ennio. "The Endogenization of Technological Change: A Development Challenge." In Sunkel, 1993, pp. 223-253.

Romer, Paul M. "Increasing Returns and Long-Run Growth." *Journal of Political Employment* 95:5(October 1986): 1002-1037.

________. "Endogenous Technological Change." *Journal of Political Economy* 98 (1990): S71-S102.

Rosales, Osvaldo. "An Assessment of the Structuralist Paradigm for Latin American Development and the Prospects for its Innovation." *CEPAL Review* No. 34(April 1988): 19-36.

Sabato, Jorge A. "The Influence of Indigenous Research and Development Efforts on The Industrialization of Developing Countries." In *Industrialization and Development,* H. E. Hoelscher and M. C. Hawk, eds., pp. 178-183. San Francisco: San Francisco Press, 1969.

Sagasti, Francisco. *Science and Technology for Development: Main Comparative Report of the Science and Technology Policy Instruments Project.* Ottawa: International Development Research Centre, 1978.

Salazar-Xirinachs, Jose Manuel. "The Role of the State and the Market in Economic Development." In Sunkel, 1993, pp. 361-395.

Schwartzman, Simon. "Coming Full Circle: A Reappraisal of University Research in Latin America." *Minerva* 24 (Winter 1986): 456-475.

________. "The Power of Technology." *Latin American Research Review* 24:1 (1989): 208-221.

________. *A Space for Science: The Development of the Scientific Community in Brazil.* University Park: Pennsylvania State Press, 1991.

Sercovich, Francisco C. *Technologia and Control Extranjeros en la Industria Argentina.* Buenos Aires: Siglo Veintiuno, 1975.

Sosa de Golindano, Graciela and Jesus Pena Cedillo. "Venezuela." In Oro y Sebastian, pp. 379-410.

Soto Angli, Francisco and Victor Castillo. *El Economia Politica de la Interrelacion Technologica entre Mexico y los Estados Unidos: El Caso de la Industria Petroquimica.* La Jolla, CA: Center for U.S.-Mexican Studies, University of California, San Diego, 1983.

Strassmann, W. Paul. "Employment and Financial Alternatives in Mexican Housing." In *Studies on Employment in the Mexican Housing Industry,* Christian Araud et al., eds., pp. 227-303. Paris: Organization for Economic Cooperation and Development, 1973.

________. "The Development of Alternative Construction Technologies in Latin America." In *Technological Progress in Latin America: The Prospects for Overcoming Dependency* edited by James H. Street and Dilmus D. James, pp. 247-257. Boulder, CO: Westview 1979.

________. "The Rise, Fall, and Transformation of Overseas Construction Contracting," *World Development* 17:6 (June 1989): 783-794.

Street, James H. and Dilmus D. James. "Institutionalism, Structuralism and Dependency in Latin America." *Journal of Economic Issues* 16:4 (September 1982): 673-698.

Sunkel, Osvaldo. "Underdevelopment, the Transfer of Science and Technology in the Latin American University." *Human Relations* 24:1 (1971): 1-18.

________. "Structuralism, Dependency and Institutionalism: An Exploration of Common Ground and Disparities." *Journal of Economic Issues* 23:2 (June 1989): 519-533.

________. "From Inward-looking Development to Development from Within." In Sunkel, 1993, pp. 23-59.

________ and Joseph Ramos. "Toward a Neostructuralist Synthesis." In Sunkel, 1993, pp. 5-19.

________ and Gustavo Zuleta. "Neo-Structuralism versus Neo-Liberalism in the 1990s." *CEPAL Review* 42 (December 1990): 35-51.

________. ed. *Development from Within: Towards a Neostruturalist Approach for Latin America.* Boulder, CO: Lynne Rienner, 1993.

Swaney, James A. and Robert Premus. "Modern Empiricism and Quantum-Leap Theorizing in Economics." In *Why Economics in Not Yet a Science* edited by Alfred S. Eichner, pp. 41-60. Armonk, NY: M. E. Sharpe, 1983.

Tokman, Victor. "Distribution of Income, Technology and Employment: An Analysis of the Industrial Sectors of Ecuador, Peru, and Venezuela." *World Development* 2:10/11/12 (October/November/December 1974): 49-58.

________. *Employment and Technological Change in Capital Goods Production in Latin America.* Santiago: PREALC, 1981.

________. "Labor Markets and Employment in Latin American Economic Thinking." In Sunkel, 1993, pp. 127-149.

Unger, Kurt and Viviane Marquez. *La Tecnologia en la Industria Alamentaria Mexicana.* Mexico, DF: El Colegio de Mexico, 1981.

Vaitsos, Constantine. *Intercountry Income Distribution and Transnational Enterprises.* Oxford: Clarendon Press, 1974.

Vessuri, Hebe. *Las Instituciones Cientificas en la Historia de la Ciencia en Venezuela.* Caracas: Fondo Editorial Acta Cientifica Venezolana, 1987.

________. "Higher Education, Science and Engineering in Late 20th Century Latin America: Needs and Opportunities for Co-Operation." *European Journal of Education* 28:1 (1993): 49-49.

Villaschi, Arlindo. *The Newly Industrialized Countries and the Information Technology Revolution: The Brazilian Experience.* Aldershot: Avebury, 1994.

Wilson, Steven R., Robert Ballance and Janos Pogany. *Beyond Quality: An Agenda for Improving Manufacturing Capabilities in Developing Countries.* Aldershot: Edward Elgar, 1995.

Wionczek, Miguel. *Inversion y Tecnologia Extranjera en America Latina.* Mexico, DF: Joaquin Mortiz, 1971.

________ and Miguel H. Marquez. "Mexico." *In Technology Transformation in the Third World: Latin America*, Surendra J. Patel, ed., Vol. III., pp. 1-100. Aldershot: Averbury, 1992.

8

SOME POSSIBLE DIFFERENCES BETWEEN AMERICAN AND EUROPEAN INSTITUTIONALISM

Geoffrey M. Hodgson[1]
University of Cambridge

Institutional economics in America has a long and vibrant history. It can date its birth to 1898, when Thorstein Veblen published his seminal essay "Why is Economics not an Evolutionary Science?" in the *Quarterly Journal of Economics*. By the 1920s, institutionalism was virtually dominant as a school of thought in American universities. In contrast, in Europe, institutionalism has never been firmly established in the twentieth century, despite the important contributions of major European institutionalists such as John Hobson, Karl Polanyi, Gunnar Myrdal and William Kapp. European institutionalism has a fragmented history, lacking in strong continuity, unlike the clear generational linkages in America linking Thorstein Veblen, John Commons and Wesley Mitchell with successors such Clarence Ayres, Allan Gruchy and John Kenneth Galbraith.

Walrasian, Marshallian, Keynesian and other assorted ideas spread from the Old World to the New, but there was very little traffic in institutionalism. By the time the United States became a prominent exporter of economic theory - that is after the Second World War - institutionalism had been displaced from pre-eminence within that country by the now familiar formalisms of neoclassical economics.

It is only recently that a European institutionalism has begun to emerge, with its own organizational vehicle. At the time of writing there are two continental-based associations of institutional economics. In North America the Association for Evolutionary Economics was formed about thirty years ago. In Europe the European Association for Evolutionary Political Economy was formed much later, in 1988. Yet their memberships are now roughly equal in numbers, with AFEE at just less than 600 and EAEPE at just over 700. Comparisons with the respective orthodox economics associations in the two continents are even more graphic. The American Economic Association has about 17,000 members in North America. Accordingly, the ratio of the mainstream to institutionalism in America is roughly 30:1. The European Economic Association has about 1,400 members in Europe.[2] Hence the membership ratio of the mainstream to institutionalism in Europe is about 2:1. Although these measures are crude, they serve to dramatize the rising comparative potential of institutional economics in Europe.

Can we go so far as to suggest that there is an eastwards shift of the center of gravity of institutionalism, back towards Europe, roughly 100 years after the son of a Norwegian immigrant penned his famous essay in Chicago and placed a new school of economic thought on the New World map?

This essay considers the possibility of the emergence of a distinctive version of European institutionalism in the twenty-first century. It is reasonable, first, to examine the historical background. In particular, it is necessary to address the question: Why did the American institutionalism of Veblen, Commons, and Mitchell fail to spread widely in Europe in the first half of the twentieth century? This question is addressed in Part I, which also considers the wider intellectual contexts of American and European institutionalism. Part II considers possible theoretical divergences between these traditions and Part III concludes the essay.

CONTEXTS

The Sporadic Growth of European Institutionalism in the Twentieth Century

An explanation of the different growth rates of European and American institutionalism should focus, in part, on the differences in academic culture and academic institutions in two continents. First, in the period from 1890 to 1930, when institutionalism established itself in America, the United States was enjoying a rapid increase in population and a corresponding growth in the number of universities. There was an excess demand for qualified academics. Consequently, there was an expanding academic space for heterodoxy in economics to gain a foothold in the universities.

Second, although socialist ideas did spread in the United States before the First World War, Marxism has never been as prominent there as in Europe. Marxian social science did not develop much in America, compared with, for example, in Germany in the 1890-1930 period and in Britain in the 1930s. Accordingly, in the United States, institutionalism was not rivaled by this alternative heterodoxy to the same degree. The weakness of Marxism in America provided more space for institutionalism.

Third, the United States was a new and expanding country in which the very problems of institutional and organizational construction and long-term development were immediate, problematic and matters for deliberation within the social culture. Although Continental Europe has endured many political convulsions in the nineteenth and twentieth centuries, the country in which much economic science had developed since Adam Smith was, by contrast, stable and relatively ossified. In

Britain, institutions which had seemingly endured for centuries could apparently be taken for granted. There it would seem safe to ignore problems of long-term development and change because the safest prediction of the future was that it would be very much the same as the present. If so, the focus on cumulative development and institutional change in institutionalist theory would have seemed less relevant to a British economist, even if such complacency was in retrospect misconceived.

There may be other reasons why institutionalist ideas did not spread to Europe, such as the theoretical limitations of American institutionalism itself. It has been widely accepted, both by supporters and critics of institutionalism, that it has not offered a theoretical system as comprehensive or coherent as, say, neoclassicism or Marxism. There are big differences between the ideas and approaches of its leading figures. The heterogeneity of the school provided a fruitful and productive pluralism within the United States, but it was not a well-defined and attractively packaged item for export.

Fragments of European Institutionalism

The reasons outlined above may partly explain the failure of an established school or network of institutionalists to emerge in Europe prior to 1988. However, there are several important precursors on the European scene. They are not readily identifiable because the institutionalist label was never widely adopted in Europe, and institutionalism itself contains a diversity of ideas. The task of identifying institutionalist theory in the Old World amounts to the identification of the few sympathetic personages and similar schools of thought.

One of the most important resonances is with the German historical school of the nineteenth and early twentieth century, including Bruno Hildebrand, Karl Knies, Wilhelm Roscher, Gustav Schmoller, Werner Sombart and others. Notably, early precursors of American institutionalism such as Richard Ely and Edwin Seligman studied economics in Germany in the 1870s and 1880s and under the tutelage of leading members of the historical school. However, the ground tilled by that school in Germany did not become fertile ground for an institutionalism of the American type, partly because the school lost out in the famous Methodenstreit with Carl Menger, and partly because Germany and Austria were subsequently overwhelmed by Nazism in the 1930s. In addition, the important theoretical differences between institutionalism and the historical school should not be ignored.[3]

Although it was the undisputed world center of economic science from 1776 to 1945, Britain seems to have been particularly infertile ground for institutionalism, with few figures with explicit or even implicit institutionalist leanings. This is despite the influence of major British thinkers such as Charles Darwin and Herbert Spencer on Thorstein Veblen. In the period up to the Second World War the most important British economist who could be given the institutionalist label is John A. Hobson. Notably, Hobson never occupied an academic post at a British university and was never accorded significant recognition by the British academic establishment. Nevertheless, his ideas influenced thinkers as significant as Vladimir Ilyich Lenin and John Maynard Keynes.

It is in Continental Europe that we find the three major figures with the closest and most explicit links with American institutionalism, namely K. William Kapp (Germany and Switzerland), Gunnar Myrdal (Sweden) and Karl Polanyi (Austria and Hungary). All three have made major contributions to institutional thought, of broadly equivalent significance to those of Veblen, Commons or Mitchell in America.

More specifically, Kapp has made major contributions to the theory of social cost and the environmental impact of business enterprise, as well to general institutional theory. Myrdal is well known for his contribution to monetary theory, as well as his later applications of the theory of cumulative causation to studies of racial disadvantage, regional disparity and third world underdevelopment. Polanyi made highly significant contributions to economic history, economic anthropology and institutional theory. He argued that the development of the capitalist market system during the industrial revolution was not and could not be entirely spontaneous but necessarily involved the legislation, regulation and meddling of the state.

However, despite their contributions, none of these figures has left behind a thriving school of disciples with significant influence in the European universities. Strangely, it is outside their home countries that they are most recognized. In the case of Myrdal, for example, his ideas on cumulative causation were taken up most prominently by Nicholas Kaldor at the University of Cambridge in England. Polanyi eventually immigrated to the United States and established an important following there. Kapp spent most of his adult life teaching in the United States and in Switzerland.

An European Context for Institutionalism

Institutional economics can be defined, quite broadly, in terms of:

(a) the rejection of individualistic assumptions of hedonism and exogenous preferences in favor of an organicist conception of the individual;

(b) the adoption of institutions as a major unit of analysis rather than an exclusive focus on individuals;

(c) the treatment of the economy as an open rather than a closed system;

(d) the rejection of an exclusive emphasis on equilibrium in favor of the idea of cumulative causation.

In these terms there are a number of other European figures who come close to institutionalism in their theoretical writings. They include very prominent economists such as Karl Marx and Joseph Schumpeter. In addition there are elements of institutionalist thinking in the works of John Maynard Keynes, and even in those of Adam Smith, Alfred Marshall and Friedrich Hayek.

Consequently, if a revitalized institutional economics is to emerge in Europe it is likely to draw on some or all of these alternative sources of inspiration. Intellectual traditions inspired by Marx and Keynes will be of particular importance, including modern developments as Post Keynesianism, and strains of Marxism such as that associated with the French régulation school. There are also resonances with some of the ideas of Hayek, especially his analysis of the evolution of rules and his stress on the importance of knowledge, particularly if this analysis can be freed from its conventional associations with a strident, free-market ideology.

The conditions for a European revival of institutionalism have improved in recent years, partly because Europe is undergoing major institutional, cultural and technological changes. It has endured many of these in the past, but often such convulsions have been associated with totalitarianism and the destruction of much academic life. In the 1990s, however, enormous changes are occurring in Eastern Europe along with processes of political and economic integration in the West, and without any major threat to academic liberties.

These developments offer an opportunity for institutional economics, in part because of the failures of neoclassical orthodoxy. Neoclassical theory normally ignores questions of technological change and has a limited and stunted theory of institutions in which individuals are taken as given. The emphasis within institutionalism on the moral and ethical dimensions of economics, along with its attention to matters of cultural and technological development, gives it a significant practical and theoretical advantage.

Another reason why the "old" institutional economics may enjoy a revival in Europe is due to differences in corporate and political culture. Modern Continental Europe presents "corporatist" and "social market" versions of capitalism, in contrast to the ideology (but perhaps not so much the corporate reality) of the more individualistic "casino economy" of the United States. Through European integration, Britain too is being forced to conform more and more to Continental European norms and to pull away from the traditional individualism that it has shared historically with America. With its emphasis on the social and institutional nature and context of all human actions, the "old" institutionalism may become more and more at home in the European arena, and perhaps also while neoclassical economics continues to thrive in the more individualistic ideological environment of the United States.

DIVERGENCE?

Possible Differences Between European and American Institutionalism

Despite the contextual differences, few arguments have yet been given here to suggest that European institutionalism might develop along lines different from its American counterpart. Indeed, all three of the most prominent European institutionalists - Myrdal, Polanyi and Kapp - spent a much of their lives in the United States, and were thus influenced by leading American institutionalists of their time. Even the stronger relative influence of Marxism in Europe does not necessarily portend a wide transatlantic divergence in institutionalism, as a number of the younger, post-1968 institutionalists in America exhibit equivalent affinities to elements of Marxian theory. Indeed, there is a vocal group of American "radical institutionalists" attempting a synthesis between elements of institutionalism and Marxism (Dugger 1989; Stanfield 1995). It could thus be suggested that convergence rather than divergence is likely.

However, before deciding on this issue, there other issues to be considered. Arguably, more important sources of potential divergence lie elsewhere, in part reflecting different philosophical and cultural traditions. I shall locate these in terms of two quite different approaches to the theory of valuation and welfare, emanating from different emphases on the nature and role of institutions. In addition, it shall be argued that the histories of both American and European institutionalism are path-dependent, and the traumatic decline of institutionalism in America after 1930 disposed it to a confined pattern of subsequent theoretical development that is unlikely to be paralleled in Europe.

Institutions as Constraints or Enablers?

American social science was developed in an ideological context that emphasized individuality and free choice. By the 1920s, earlier arguments that human behavior was significantly influenced by biological instincts became extremely unpopular in American academia (Degler 1991). For this reason, American institutionalists ended their reliance on the instinct psychology of William James and others. This meant dispensing with the idea of instincts and a major component of their Veblenian intellectual heritage (Hodgson, forthcoming). However, Veblen was not a biological determinist. For him, instincts were molded and sometimes even engendered by institutions (Rutherford 1984). This is clearly consistent with the view that institutions are constitutive of action as well as providing constraints upon it (Giddens 1984; Hodgson 1988).

The idea of the dual role of institutions as both enablers and constraints survived in the writings of Commons (1934, 73), who clearly saw institutions as a liberating as well as a constraining force. But the clear vision of dual enabling and constraining roles was not retained. This is especially the case in the post-1945 period, when Ayres' influence on American institutionalism gave it a new and distinctive flavor. In the writings of Ayres, and in accord with the prevailing libertarian ideology of American society, the greater stress is on the conventional idea of an institution as a constraint (McFarland 1985, 1986).

Ayres saw an institution largely as a negative barrier to progress, as "a bad thing from which we are bound to try perpetually to redeem ourselves" (Ayres to Dewey, 29 January 1930, quoted in Tilman 1990, 966). Ayres did not see institutions as providing essential coordination and cognitive guidance, but as ceremonial or archaic vestiges of a more primitive past. Hence he wrote: "The history of the human race is that of a perpetual opposition of these forces, the dynamic force of technology continually making for change, and the static force of ceremony-status, mores, and legendary belief opposing change" (Ayres 1944, 176). This alleged tension between dynamic technology and static institutions was dubbed the "Veblenian dichotomy" (Waller 1982). However, a dichotomy between "ceremonial" and "instrumental" institutions is not so clearly apparent in Veblen's writings as Ayres and his followers presume.[4]

Apart from the doubts concerning precedence in Veblen's own work, the very idea of a dichotomy between institutions and technology has been challenged (Samuels 1987; Klein 1995). Arguably, technology itself is made up of institutionalized routines (Dosi 1988). If so, the dichotomy is invalid on logical grounds. Furthermore, particularly in an ongoing and dynamic framework, constraints are often themselves enabling (Hodgson 1988; Kenworthy 1995). Although it would be wrong to see the different views on the constraining or enabling nature of institutions as neatly categorized into "American" and "European" positions, Ayres's rather extreme and one-sided stress on the constraining function of institutions has given post-1945 American institutionalism a distinctive hallmark that is not paralleled in the European context. These differences are also relevant to the questions of valuation and welfare, to which we now turn.[5]

Valuation and Welfare in American Institutionalism

In both continents, institutionalism faces the problem of developing an alternative theory of valuation and welfare, to rival the ubiquitous and utilitarian approach of neoclassical economics. The founders of American institutionalism - Veblen, Commons and Mitchell - all rejected the hedonistic "lightning calculator of pleasures and pains" (Veblen 1919, 73) that characterized neoclassical theory. They rejected a theory of valuation and welfare based on individual utility.

What did they put in its place? Herein lies a long tale that we can only touch on here. Although Veblen's work was laced with satire and evaluative scorn, he did little more than hint at the foundations of an alternative approach. Institutionalism was already over a quarter of a century old when Commons (1924, 356-9; 1934, 649-875) formulated a theory of "reasonable value."

Subsequently, Ayres (1944, 1952) attempted to develop an "instrumental value theory" for institutional economics, attempting to synthesize the philosophy of John Dewey with the so-called "Veblenian dichotomy." Thus there is an attempt to develop a "theory of instrumental value" and establish a dynamic "social value principle" which "provides for the continuity of human life and the noninvidious recreation of community through the instrumental use of knowledge" (Tool 1979, 291).[6]

It is not the place to evaluate fully the Veblen-Dewey-Ayres approach here, but the following brief points must be noted. First, following Yngve Ramstad (1989), the Veblen-Dewey-Ayres theory of instrumental value is not strictly compatible with Commons' theory of reasonable value. Second, the operational utility of the instrumentalist approach is itself open to question. As Ramstad (1989, 766) puts it:

> I have frankly not found it easy to grasp how, in the face of complex circumstances, instrumental value theory provides a "normative compass" sufficient for the analyst to choose between alternative institutional adjustments, that is, what "instrumental efficiency" might usually mean with respect to the complex tradeoffs that actually exist in a specific case, other than whatever the analyst him/herself considers to be "instrumental."

Despite these lingering problems and controversies, the influence of Ayres on postwar American institutionalism has been immense. Marc Tool (1994, 16) noted: "Ayres and his students have been among the most significant contributors to the development of institutional economics in the last half-century." However, in sharp contrast, Ayres' influence on European institutionalism has been negligible. Even Dewey, one of the giants of twentieth century Western philosophy, is overall less prominent in the European intellectual scene.

Furthermore, the fragmented history of European institutionalism has meant that neither Veblen nor Commons have received the amount of attention that they have received in North America. Neither Commonsian "reasonable value" nor Ayresian "instrumental value" have as yet become rooted in European soil. It is thus in the sphere of valuation and welfare that the divergences between American and European institutionalism are at their widest.

Valuation and Welfare in European Institutionalism

What can pose as a European alternative on questions of valuation and welfare? Despite the incompleteness of this endeavor, Europeans have over two thousand years of intellectual history on which to build. Indeed a central and persistent dichotomy in modern heterodox economics goes back to Aristotle. In his *Politics* (Book I, chs. 8-10) can be found the famous distinction between use-value and exchange-value, that sometimes reappears, in unacknowledged and mangled form, in some of the many derivative versions and variants of the alleged "Veblenian dichotomy."

Of course, the Aristotelian dichotomy between value in exchange and value in use was taken up by the classical economists Adam Smith and David Ricardo, and by Karl Marx. In the twentieth century it was adopted by the European institutionalist, Karl Polanyi, in his challenge to the subjectivist and utilitarian foundations of economic orthodoxy (Stanfield 1986). Indeed, Polanyi's interpretation of these concepts was closer to Aristotle than some modern interpretations that have wrongly identified use-value with subjective utility. For Polanyi, as for Aristotle, use-value had an objective quality relating to the usefulness of an item for humankind. Its pursuit was more obvious in traditional institutions such as the household, and eclipsed by the quest for monetary gain in the market.

Opposition to the reduction of all human values to a single calculus of money or utility has persisted for centuries (Lutz and Lux 1979, 1988). One of the notable and influential European critics of utilitarianism and of the pecuniary calculus is John Ruskin. In his book *Unto This Last* (1866, 17) he attacks the idea, promoted by economists then and now, that the social affections "are accidental and disturbing elements in human nature; but avarice and the desire of progress are constant elements." For him, faith, generosity, honesty and self-sacrifice were just as real and potentially durable. Furthermore, he argued that economic policies that appealed simply to hedonistic motives are ultimately self-defeating because they undermined morality and social cohesion.

The British institutionalist Hobson was influenced by Ruskin and saw the importance of appeals to moral values in economic and social policy. For example, in his discussions of motives and incentives in the workplace, Hobson (1914, 302-4; see Freeden 1988, 82-5) places emphasis on "organic" moral values and on the limitations of atomistic self-interest. Hobson combines the basic "old" institutionalist tenet of the malleability of human purposes and preferences - prominent from Veblen to Galbraith - with an assertion of the centrality of social norms and moral values to economic policy. At the same time, Hobson insists on the importance of material and pecuniary incentives to motivate individuals. To some degree economic policy has to appeal to self-interest, but at the same time it has to transcend it by considering the "organic" interests of society as a whole.

Ruskin's attack on hedonism and Benthamite utilitarianism resonates through British culture. In such a spirit John Maynard Keynes (1933, 445-6) wrote:

> I do now regard that [Benthamite] tradition as the worm which has been gnawing the insides of civilization and is responsible for the present moral decay. ... We used to regard the Christians as the enemy, because they appeared as the representatives of tradition, convention and hocus pocus. In truth it was the Benthamite calculus based on an over-valuation of the economic criterion, which was destroying the quality of the popular ideal.

An equally strong critique of utilitarianism emanates from the Durkheimian sociological tradition in France. Durkheim argued that there is no direct and necessary correlation between human happiness and economic prosperity. His emphasis was on the need for moral, as well as political and economic, order in society. Society requires moral as well as technocratic regulation. Hence conceptions of "socialism" that were built on some self-evident, rational economic ideal, without recourse to a pervasive moralization of socio-economic life, were misconceived and doomed to failure (Durkheim 1972).

It is not suggested that the ideas of Ruskin or Durkheim are adequate for our task. It is simply proposed that such thinkers constitute a tradition quite different from that which pervades American social science. This tradition stresses moral considerations that are not necessarily rooted in material prosperity, correlated with technological advance, nor subsumed under the category of individual gratification or liberty.

Modern neoclassical economics clearly bears traces of the prolonged influence of classical liberalism. Alasdair MacIntyre (1985) bemoans the liberal culture in which moral values are rendered relative and subjective, the gratification of emotions is seen as paramount, and the notion of the good is deemed to be a purely private matter. Each person is declared free to judge the moral good for herself, and is deemed to be at liberty to pursue that private vision subject to the constraint that it does not impinge on those like freedoms for others. The abandonment of the search for a greater "common good" has tragically been proclaimed as the era of freedom and human liberation. As Brian Crowley (1987, 18) argues: "Having dispensed with any claim to be able to judge the goodness of the ends that men pursue, the moral interest of society in individual or collective action turns from the ends to be achieved to focus exclusively on the means chosen to achieve them."

However, while rejecting much of the subjectivism and individualism of the liberal and neoclassical creeds, American institutionalism may not yet go far enough in establishing an independent moral discourse concerning the interests of society and humanity. It has sometimes fallen back on scientism and technophilia, mirrored in Dewey's excessive faith in scientific evaluation through experiment (Scheffler 1974) and in Ayres' "gospel of technology" (McFarland 1986; Rutherford 1981). It is in the statements of European thinkers such as Ruskin and Durkheim that we have the clearest insistence on the need for a moral discourse that is not subservient to such exigencies.

So far, the work of European institutionalists in this area is underdeveloped. However, there is a clear tradition of analysis that emphasizes levels of welfare discourse that transcend both scientism and subjectivism. Examples include the pioneering work of Kapp (1978) on environmental policy and social cost, and the recent work by Len Doyal and Ian Gough (1991) on the theory of human need.

Path Dependence in the Histories of American and European Institutionalism

Ever since Veblen, institutionalists have rightly emphasized the existence of cumulative causation and path-dependence. It is reasonable, therefore, that such ideas should be applied to the history of institutionalism itself. Elsewhere (Hodgson, forthcoming) I have considered the history of American institutionalism since the 1890s and some of the reasons for its decline after the 1930s. In part this was due to the reaction against biological ideas in social science and against the use of evolutionary metaphors (Degler 1991). Following the "golden age" of the late nineteenth century, the inter-war period was the "dark age" for evolutionism in social science: "even the word 'evolution' came to be uttered at serious risk to one's intellectual reputation" (Sanderson 1990, 2).

Veblen's project to build a "post-Darwinian" (Veblen 1919, 37, 192n.) evolutionary economics was hindered by problems in the scientific development of Darwinian biology from 1900 to 1940. Having failed to develop a theoretical system to rival neoclassicism or Marxism, in the changing intellectual context institutionalism was insecure. The shift away from "economic biology" thus helped the marginalization of institutionalism and the triumph of neoclassicism.

Institutionalism was also vulnerable to shifts in the prevailing conception of scientific methodology. From the turn of the century, and in the name of science, strong American voices argued for reductionism and methodological individualism in social analysis. The shift in thinking towards reductionism was facilitated by the reduction in popularity of organic analogies taken from biology. Reductionist notions - claiming the spurious authority of "scientific methodology" - did much damage to the institutionalists. John Commons seemed especially vulnerable with his concept of "collective action."

Around 1900, the pragmatist ideas of Charles Sanders Peirce and William James had been formative for institutionalism in America. In the 1920s, both the triumph of positivism and the growing unpopularity of instinct psychology struck at the pragmatist foundations of institutional economics. The rise of positivism meant that the Peircian methodological project to transcend both deduction and induction was pushed to one side. Peircian and other metaphysical and ontological speculations became unfashionable, to be replaced by a naive faith in the unaided authority of evidence and experiment. The precipitous decline of instinct psychology also created severe difficulties for institutionalism. Deprived of such psychological foundations, the institutionalist critique of the rational actor paradigm was traumatized and arguably weakened. Considering this onslaught against its core ideas it is amazing that the institutionalism of Veblen and Commons survived as long as it did.

American institutionalism survived the inter-war period by downplaying part of its legacy. This is best illustrated by considering the later generation of institutionalists, educated in the inter-war years and rising to prominence after the Second World War. As we have noted, the leading and by far the most influential American institutionalist in this category is Ayres. Largely at Ayres' instigation, the institutionalists shifted their philosophical emphasis from Peirce to Dewey (Mirowski 1987). Dewey had greater faith that science in general, and scientific experiment in particular, could serve both as routes to truth and as the bases of social policy. Although Dewey was not a positivist and was a critic of many of the inter-war developments in scientific methodology, his reverence for science fitted in well with the times.

The positivistic climate of the 1920s pushed the institutionalists towards empiricism. Mitchell, one of the most influential living exponents of institutionalism, became increasingly engrossed in statistical studies. The statistical turn was resisted by others, "arguing that statistics offered a useful empirical and analytical tool but could not remake a theory" (Ross 1991, 415). But Mitchell was resolute. According to Dorothy Ross, these debates in the late 1920s were a "turning point" in the fortunes of institutionalism, evincing a gathering impatience with the failure to develop a systematic theory.[7]

The Great Crash of 1929 and the subsequent depression brought further difficulties. The personal recollections of Gunnar Myrdal are particularly apposite. When he came to the US at the end of the twenties, institutional economics was still seen by many as the "wind of the future." However, at that time Myrdal was "utterly critical" of this orientation in economics and himself at the "theoretical" stage of his own development. Myrdal goes on to explain a key event in the decline of the popularity of institutionalism in the United States:

> What I believe nipped it in the bud was the world-wide economic depression. Faced with this great calamity, we economists of the "theoretical" school, accustomed to reason in terms of simplified macro-models, felt we were on the top of the situation, while the institutionalists were left in a muddle. It was at this stage that economists in the stream of the Keynesian revolution adjusted their theoretical models to the needs of the time, which gave victory much more broadly to our "theoretical" approach (Myrdal 1972, 7)[8]

Without referring to Myrdal, Ross (1991, 419) corroborates the argument:

> Institutionalism as a movement ... fell victim to the Great Depression and its Keynesian remedy. For self-proclaimed experts in historical change, their inability to come to any better understanding of the Depression than their neoclassical colleagues was a considerable deficit. Mitchell in particular, who predicted like everyone else that the downturn would right itself within a year or two, was driven deeper into his program of empirical research by this proof of ignorance. Whether a more powerful and genuinely historical institutional economics would have done better is impossible to say. Like the left-liberal economists generally, the institutionalists were drawn into the Keynesian revision of neoclassicism.

Of course, the rising "Keynesianism" of the 1930s was different in several key respects from the economics of Keynes. Keynes himself was at best skeptical of econometrics and mathematical modeling in economics. What did emerge in the 1930s were the foundations of the neoclassical-Keynesian synthesis, based on key developments in neoclassical microeconomics and a mechanistic system of macroeconomic modeling with some Keynesian affinities.

Institutional economists were far from hostile to the Keynesian revolution. Indeed, the rise of Keynesianism coincided with a drift of Anglo-American opinion towards state intervention and planning, and in the 1930s American institutionalists were active in the inspiration, development and promotion of President Roosevelt's New Deal (Barber 1994; Stoneman 1979; Wunderlin 1992). Institutionalism formed its own synthesis with Keynesianism, giving less emphasis to mathematical modeling and interpreting Keynes in terms of an organicist ontology (Dillard 1948; Gruchy 1948).

In summary, therefore, American institutionalism survived the intellectual shifts of the first three decades of the twentieth century and the shock of the Great Crash of 1929 by making some fundamental adjustments in its assumptions and research program:

- Veblen's (1898) explicit project of building an evolutionary economics based on "post-Darwinian" principles taken from biology (Hodgson 1993, ch. 9) was quietly abandoned by the 1920s. Close connections between biology and the social sciences had become unfashionable. The term "evolutionary economics" was retained by the American institutionalists but largely evacuated of any biological connotations.

- By the 1920s Veblen's behavioral theory built on James's instinct psychology was marginalized. Accordingly, the concept of instinct was downplayed but the notion of habit retained. American institutionalism thus failed to develop a sufficiently distinctive theory of human agency and relied increasingly on compatible notions of enculturation taken from anthropology and sociology.

- These theoretical adjustments derailed the institutionalist research program in its infancy. To compensate, and in accord with a strong positivist drift in social science, increasing relative emphasis was put on "data gathering" and statistical work.

These pre-1940 developments in American institutionalism contributed to its distinctive character, and especially in the declining years of its popularity after the Second World War. According to the Veblenian pattern: "The situation of today shapes the institutions of tomorrow through a selective, coercive process, by acting upon men's habitual view of things, and so altering or fortifying a point of view or a mental attitude handed down from the past" (Veblen 1899, 190).

It is thus not surprising that modern American institutionalists do not link themselves, however critically, with modern applications of the evolutionary paradigm to economics notably by Richard Nelson and Sidney Winter (1982). They make little use of the burgeoning developments in evolutionary theory in biology and their applications to the social sciences. Against the spirit of Veblen, the biologization of economics is often resisted (De Bresson 1987; Jennings and Waller 1995). As the word "evolutionary" appears with increasing frequency in varied quarters of economics (Hodgson 1995), and is associated with a new wave of "economic biology", its use by the American institutionalists seems more and more an anomaly.

By contrast, European institutionalism is presently in a more fluid state, and its boundaries with other heterodoxies are less clear. Not only is the influence of Marxian economics greater, there has also been a sustained dialogue with Schumpeterians, the French régulation school and evolutionary theorists influenced by Nelson and Winter. Note in particular the work of Giovanni Dosi and his collaborators in Italy, and the research of Christopher Freeman and others at the Science Policy Research Unit at the University of Sussex in the UK. Together with Nelson and others, Dosi and Freeman are co-authors of a major work on technical change (Dosi et al 1988). There have been fruitful applications of these ideas, particularly in the sphere of technology policy, corporate strategy and national systems of innovation. The European Association for Evolutionary Political Economy is strongly networked into all these developments.

Accordingly, the term "evolutionary economics" has stronger resonances in the European institutionalist arena than it does among institutionalists in America. Even if, according to the Veblenian scenario, European institutionalism becomes more ossified, its likely channels of development will show significant differences from those of postwar American institutionalism.

CONCLUDING REMARKS

We must never ignore the historical accidents and idiosyncrasies in the acquisition of labels. If institutionalism had had a strong base in Britain by the 1920s it is quite likely that Keynes would have called himself an institutionalist. As well as his affinity for Commons[9] he gave a long acknowledgment in the *General Theory* of his debt to J. A. Hobson - the nearest thing to a British institutionalist and someone who wrote a book on Veblen. Further, if US institutionalism had not declined so dramatically in the 1930-65 period then it is quite possible that Sidney Weintraub, Alfred Eichner and Paul Davidson would have described themselves as institutionalists. Americans typically exacerbate their lack of a long history by forgetting even the recent past.

This essay is not intended as a polemic against American institutionalists, to many of whom creative heterodox economists owe an immense intellectual and moral debt. Instead, a central objective has been to apply an institutionalist perspective on institutional development to institutionalism itself, and in particular to its two traditions, on each side of the Atlantic. Neither should it be suggested that each continent-bound version of institutionalism presents a homogenous, consistent or unified body of doctrine. There is a potentially healthy degree of diversity within both Europe and America.

Nevertheless, specific intellectual traditions and habits of thought prevail. Furthermore, it is possible to give a tentative identification of distinct continental characteristics. This is especially tentative in the European case because a common academic language in modern Europe is a very recent phenomenon. Since the decline of Latin as the common European language of academia, over two centuries have passed where French, German and English have rivaled each other as pre-eminent languages of science. It is only in the last quarter of the twentieth century that a decisive shift in favor of English as the European academic language has occurred. It is thus only recently, and by means of this development, that a distinctively European scientific identity has been possible. But the very choice of English as the common language has brought a renewed threat of American domination in science.

Concerning institutionalism, the attempt here to identify a "European" variant is no more than tentative. In any case it is hoped that fruitful dialogue between institutionalists in Europe and America will continue, with the eventual outcome of building a more realistic, operational and humane economic science. If a difference between continental schools of institutionalism is emerging then such a development can be healthy, and all can learn from the open dialogue that may ensue.

ENDNOTES

1 The author is most grateful to Dilmus James, Jack Mogab and Richard Nelson for helpful comments on an earlier version of this essay.

2 Europe is still in the process of unification and has no common language. In this context, it should be noted that he European Economic Association is relatively young and has not yet established itself with a strong membership, even among orthodox economists. The EEA membership data cited here is taken from Guesnerie (1995).

3 Veblen (1919, 58) wrote: "no economics is farther from being an evolutionary science than the received economics of the Historical School ... they have contented themselves with an enumeration of data and a narrative account of industrial development, and have not presumed to offer a theory of anything or to elaborate their results into a consistent body of knowledge. Any evolutionary science, on the other hand, is a close-knit body of theory. It is a theory of process, of an unfolding sequence."

4 One of the most refined expositions of this approach (Bush 1986) still fails in its attempt to find clear precedents for such ideas in the writings of Veblen. It is widely held that the "Veblenian dichotomy" is evident in Veblen (1914; esp. 25 quoted by Ayres 1944, 176) and in Veblen (1899, ch. 1; 1904; 1919, 279-323). I would like to put down a friendly challenge to all exponents of the allegedly general dichotomy between institutions and technology to cite clear and unambiguous evidence of it from Veblen's works. Personally, I have failed to discover such convincing evidence.

5 Several American institutionalists, notably Dugger (1995), have recently argued against Ayres' unqualified and optimistic view of technological progress.

6 One of the clearest recent expositions of "instrumental valuation" is in Tool (1995).

7 Myrdal (1958, 254), himself an institutional economist, saw inter-war American institutional economics as marked by a "naive empiricism."

8 What makes this personal testimony particularly striking is the fact that years later - in the 1940s - Myrdal converted to institutionalism and subsequently won the Nobel Prize in Economics for his sins.

9 Keynes wrote a letter dated 26 April 1927 letter to Commons: "Judging from limited evidence and at great distance, there seems to be no other economist with whose general way of thinking I find myself in such genuine accord." (John R. Commons Papers, State Historical Society of Wisconsin).

REFERENCES

Aristotle. *The Politics*, translated with an introduction by T. A. Sinclair. Harmondsworth: Penguin,. 1962.

Ayres, Clarence E. *The Theory of Economic Progress*, 1st edn. Chapel Hill, NC: University of North Carolina Press, 1944.

________. *The Industrial Economy* Cambridge, MA: Houghton Mifflin, 1952.

Barber, William J. "The Divergent Fates of Two Strands of 'Institutionalist' Doctrine During the New Deal Years." *History of Political Economy* 26:4 (Winter 1994): 569-87.

Bush, Paul Dale. "On the Concept of Ceremonial Encapsulation." *The Review of Institutional Thought* 3 (December 1986): 25-45.

Commons, John R. *The Legal Foundations of Capitalism.* New York: Macmillan, 1924. Reprinted New York: Augustus Kelley, 1974.

________. *Institutional Economics - Its Place in Political Economy.* New York: Macmillan, 1934. Reprinted with a new introduction by M. Rutherford. New Brunswick, NJ: Transaction, 1990.

Crowley, Brian L. *The Self, The Individual, and the Community: Liberalism in the Political Thought of F. A. Hayek and Sidney and Beatrice Webb.* Oxford: Clarendon Press, 1987.

Degler, Carl N. *In Search of Human Nature: The Decline and Revival of Darwinism in American Social Thought.* Oxford and New York: Oxford University Press, 1991.

De Bresson, C. "The Evolutionary Paradigm and the Economics of Technological Change." *Journal of Economic Issues* 21:2 (June 1987): 751-61.

Dillard, Dudley. *The Economics of John Maynard Keynes: The Theory of a Monetary Economy.* London: Crosby Lockwood, 1948.

Dosi, Giovanni. "The Sources, Procedures, and Microeconomic Effects of Innovation." *Journal of Economic Literature* 26:3 (September 1988): 1120-71. Reprinted in Freeman, Christopher (ed.) *The Economics of Innovation.* Aldershot: Edward Elgar, 1990.

________, Christopher Freeman, Richard Nelson, Gerald Silverberg and Luc Soete (eds). *Technical Change and Economic Theory.* London: Pinter, 1988.

Doyal, Leonard and Ian Gough. *A Theory of Human Need.* London: Macmillan, 1991.

Dugger, William M. (ed.). *Radical Institutionalism: Contemporary Voices.* Westport, CT: Greenwood Press, 1989.

________. "Veblenian Institutionalism: The Changing Concepts of Inquiry." *Journal of Economic Issues* 29:4 (December 1995): 1013-27.

Durkheim, Emile. *Selected Writings*, edited with an introduction by Anthony Giddens. Cambridge: Cambridge University Press, 1972.

Freeden, Michael (ed.). *J. A. Hobson: A Reader.* London and Boston: Unwin Hyman, 1988.

Giddens, Anthony. *The Constitution of Society: Outline of the Theory of Structuration.* Cambridge: Polity Press, 1984.

Gruchy, Allan G. "The Philosophical Basis of the New Keynesian Economics." *Ethics* 58:4 (July 1948): 235-44.

Guesnerie, R. "European Economic Association." *Royal Economic Society Newsletter* 88 (January 1995): 16.

Hobson, John A. *Work and Wealth: A Human Valuation.* London: Macmillan, 1914.

Hodgson, Geoffrey M. *Economics and Institutions: A Manifesto for a Modern Institutional Economics.* Cambridge and Philadelphia: Polity Press and University of Pennsylvania Press, 1988.

________. *Economics and Evolution: Bringing Life Back Into Economics.* Cambridge, UK and Ann Arbor, MI: Polity Press and University of Michigan Press, 1993.

________. "The Evolution of Evolutionary Economics." *Scottish Journal of Political Economy* 42:4 (November 1995): 469-88.

________. "Decomposition and Growth: Biological Metaphors in Economics from the 1880s to the 1980s." In *Evolutionary Principles of Economics,* Kurt Dopfer, ed. Cambridge: Cambridge University Press, forthcoming. To be reprinted in Hodgson, Geoffrey M. *Evolution and Institutions: Towards a Reconstruction of Economics.* Aldershot: Edward Elgar, forthcoming.

Jennings, Ann L. and William Waller. "Culture: Core Concept Reaffirmed." *Journal of Economic Issues* 29:2 (June 1995): 407-18.

Kapp, K. William. *The Social Costs of Business Enterprise*, 3rd edn. (1st edn. 1950) Nottingham: Spokesman, 1978.

Kenworthy, Lane. *In Search of National Economic Success: Balancing Competition and Cooperation.* Thousand Oaks, CA and London: Sage, 1995.

Keynes, John Maynard. *Essays in Biography.* London: Macmillan, 1933.

________. *The General Theory of Employment, Interest and Money.* London: Macmillan, 1936.

Klein, Philip A. "Ayres on Institutions - A Reconsideration." *Journal of Economic Issues* 29:4 (December 1995): 1189-96.

Lutz, Mark A. and Kenneth Lux. *The Challenge of Humanistic Economics.* Menlo Park, CA: Benjamin/Cummings, 1979.

________. *Humanistic Economics: The New Challenge.* New York: Bootstrap Press, 1988.

MacIntyre, Alasdair. *After Virtue: A Study in Moral Theory*, 2nd edn. (1st edn. 1981) London: Duckworth, 1985.

McFarland, Floyd B. "Thorstein Veblen Versus the Institutionalists." *Review of Radical Political Economics* 17:4 (Winter 1985): 95-105.

________. "Clarence Ayres and his Gospel of Technology." *History of Political Economy* 18:4 (Winter 1986): 593-613.

Mirowski, Philip. "The Philosophical Bases of Institutional Economics." *Journal of Economic Issues* 21:3 (September 1987): 1001-38. Reprinted in Mirowski, Philip. *Against Mechanism: Protecting Economics from Science.* Totowa, NJ: Rowman and Littlefield, 1988.

Myrdal, Gunnar. *Value in Social Theory.* New York: Harper, 1958.

________. *Against the Stream: Critical Essays in Economics.* New York: Pantheon Books, 1972.

Nelson, Richard R. and Sidney G Winter. *An Evolutionary Theory of Economic Change.* Cambridge, MA: Harvard University Press, 1982.

Ramstad, Yngve. "'Reasonable Value' Versus 'Instrumental Value:' Competing Paradigms in Institutional Economics." *Journal of Economic Issues* 23:3 (September 1989): 761-77.

Ross, Dorothy. *The Origins of American Social Science.* Cambridge: Cambridge University Press, 1991.

Ruskin, John. *Unto This Last: Four Essays on the First Principles of Political Economy.* London: John Wiley, 1866.

Rutherford, Malcolm C. "Clarence Ayres and the Instrumentalist Theory of Value." *Journal of Economic Issues* 15:3 (September 1981): 657-74. Reprinted in Samuels, Warren J. (ed.) *Institutional Economics,* Vol. 3. Aldershot: Edward Elgar, 1988.

________. "Thorstein Veblen and the Processes of Institutional Change." *History of Political Economy* 16:3 (Fall 1984): 331-48. Reprinted in Mark Blaug (ed.) *Thorstein Veblen (1857-1929).* Aldershot: Edward Elgar, 1992.

Samuels, Warren J. "Technology Vis-à-Vis Institutions in the JEI: A Suggested Interpretation." *Journal of Economic Issues* 11:4 (December 1977): 871-95. Reprinted in Samuels, Warren J. (ed.) *Institutional Economics,* Vol. 3. Aldershot: Edward Elgar, 1988.

Sanderson, Stephen K. *Social Evolutionism: A Critical History.* Oxford: Blackwell, 1990.

Scheffler, Israel. *Four Pragmatists: A Critical Introduction to Peirce, James, Mead, and Dewey.* London: Routledge and Kegan Paul, 1974.

Stanfield, James R. *The Economic Thought of Karl Polanyi.* London: Macmillan, 1986.

________. *Economics, Power and Culture: Essays in the Development of Radical Institutionalism.* London, Macmillan, 1995.

Stoneman, William. *A History of the Economic Analysis of the Great Depression.* New York: Garland, 1979.

Tilman, Rick. "New Light on John Dewey, Clarence Ayres, and the Development of Evolutionary Economics." *Journal of Economic Issues* 24:4 (December 1990): 963-79.

Tool, Marc R. *The Discretionary Economy.* Santa Monica, CA: Goodyear, 1979.

________. "Ayres, Clarence E." In *The Elgar Companion to Institutional and Evolutionary Economics*, Vol. 1, Geoffrey M. Hodgson, Warren J. Samuels and Marc R. Tool, eds. Aldershot: Edward Elgar, 1994, 16-22.

________. *Pricing, Valuation and Systems: Essays in Neoinstitutional Economics.* Aldershot: Edward Elgar, 1995.

Veblen, Thorstein B. "Why Is Economics Not an Evolutionary Science?" *Quarterly Journal of Economics* 12:3 (July 1898): 373-97. Reprinted in Veblen,1919.

________. *The Theory of the Leisure Class: An Economic Study of Institutions.* New York: Macmillan, 1899. Republished New York: Random House, 1961.

________. *The Theory of Business Enterprise.* New York: Charles Scribners, 1904. Reprinted by Augustus Kelley, 1975.

________. *The Instinct of Workmanship, and the State of the Industrial Arts.* New York: Augustus Kelley, 1914. Reprinted with a new introduction by M. G. Murphey and a 1964 introductory note by J. Dorfman. New Brunswick, NJ: Transaction Books, 1990.

________. *The Place of Science in Modern Civilisation and Other Essays.* New York: Huebsch, 1919. Reprinted with a new introduction by W. J. Samuels. New Brunswick, NJ: Transaction, 1990.

Waller, William J., Jr. "The Evolution of the Veblenian Dichotomy: Veblen, Hamilton, Ayres, and Foster." *Journal of Economic Issues* 16:3 (September 1982): 757-71. Reprinted in *Alternatives to Economic Orthodoxy,* R. Albeda, C. Gunn and W. Waller, eds. Armonk, NY: M. E. Sharpe, 1987; in *Thorstein Veblen (1857-1929),* Mark Blaug, ed.. Aldershot: Edward Elgar, 1992; and in. *The Economics of Institutions,* Geoffrey M. Hodgson, ed. Aldershot: Edward Elgar, 1993.

Wunderlin, Clarence E., Jr. *Visions of a New Industrial Order: Social Science and Labor Theory in America's Progressive Era.* New York: Columbia University Press, 1992.

INDEX

MIX
Papier aus verantwortungsvollen Quellen
Paper from responsible sources
FSC® C105338

If you have any concerns about our products,
you can contact us on
ProductSafety@springernature.com

In case Publisher is established outside the EU,
the EU authorized representative is:
Springer Nature Customer Service Center GmbH
Europaplatz 3, 69115 Heidelberg, Germany

Printed by Libri Plureos GmbH
in Hamburg, Germany